Volcanology: An Introduction

Volcanology: An Introduction

Edited by
Easton Nicholson

Larsen & Keller
www.larsen-keller.com

Volcanology: An Introduction
Edited by Easton Nicholson
ISBN: 978-1-63549-286-6 (Hardback)

Larsen & Keller
Published by Larsen and Keller Education,
5 Penn Plaza,
19th Floor,
New York, NY 10001, USA

Cataloging-in-Publication Data

Volcanology : an introduction / edited by Easton Nicholson.
p. cm.
Includes bibliographical references and index.
ISBN 978-1-63549-286-6
1. Volcanology. 2. Volcanoes. I. Nicholson, Easton.
QE522 .V65 2017
551.21--dc23

The publisher's policy is to use permanent paper from mills that operate a sustainable forestry policy. Furthermore, the publisher ensures that the text paper and cover boards used have met acceptable environmental accreditation standards.

Printed and bound in China.

For more information regarding Larsen and Keller Education and its products, please visit the publisher's website www.larsen-keller.com

Table of Contents

Preface

This book explores all the important aspects of volcanology in the present day scenario. It elucidates the different techniques and methods used in this subject matter. Volcanology is concerned with the detailed study of volcanoes, magma, lava and other geochemical, geological and geophysical elements related to volcanoes. It also includes the study of the formation, activity and eruptions of volcanoes. Most of the topics introduced in this text cover new techniques and the applications of volcanology. The various subfields of this discipline along with technological progress that have future implications are glanced at in great detail. This textbook is meant for students who are looking for an elaborate reference text on volcanology.

A detailed account of the significant topics covered in this book is provided below:

Chapter 1- Volcanology is the study of volcanoes, lava and magma. It deals with the formation and classification of volcanoes and also deals with the causes of these eruptions and the effect of the eruptions on other large-scale geologic processes. This chapter will provide an integrated understanding of volcanology.

Chapter 2- Volcanoes are a break in the crust of the Earth that allows lava, volcanic ash, and magma to escape from a magma chamber underneath the surface of earth. These eruptions do not only occur on the Earth but on every planetary mass object. The chapter on volcanoes offers an insightful focus, keeping in mind the theme of the chapter.

Chapter 3- Volcanoes can best be understood in confluence with the major topics listed in the following chapter. Some of the classifications of the volcanoes enumerated in this chapter are stratovolcanoes, shield volcanoes, mud volcanoes etc. Whereas stratovolcanoes erupt periodically, shield volcanoes are entirely built with magma fluid, and once shield volcanoes erupt, the lava travels farther than the lava erupted by stratovolcanoes.

Chapter 4- The major features of volcanoes covered in this chapter are fissure vent, lava dome and volcanic cone. Volcanic cones are built from the ejected particles from a volcanic vent, which subsequently piles around the vent in the shape of a cone, whereas fissure vents are linear volcanic vents through which lava erupts but usually without exploding.

Chapter 5- Sedimentary rock is one of the main types of rocks, which covers metamorphic rocks and igneous rocks. Sedimentary rocks form the deposition and cementation of the Earth's surface while the cooling and then solidification of magma or lava forms the igneous rocks. The topics discussed in the chapter are of great importance to broaden the existing knowledge on volcanic rocks.

Chapter 6- Broadly there are three types of volcanic eruptions, namely, magmatic eruptions, phreatomagmatic eruptions and phreatic eruptions. These broadly defined types have several subtypes such as vulcanian eruptions, plinian eruptions and surtseyan eruptions. Volcanology is best understood in confluence with the major topics listed in the following chapter.

Chapter 7- The allied fields of volcanology such as geology, geophysics and geochemistry are examined in this chapter. The study of solid earth and the rocks it is composed of is geology while geophysics is the science concerned with the physical properties and processes of the Earth. All these fields are mainly concerned with the physical aspects and processes of our planet.

Chapter 8- The theory that the Earth's outer shield is divided into several plates, some of these plates being large in size while some being small, is identified as plate tectonics. In plate tectonics, divergent boundary is a linear feature that exists between two tectonic plates, which are moving away from each other. This chapter will provide an integrated understanding of plate tectonics.

I would like to make a special mention of my publisher who considered me worthy of this opportunity and also supported me throughout the process. I would also like to thank the editing team at the back-end who extended their help whenever required.

Editor

1

Introduction to Volcanology

Volcanology is the study of volcanoes, lava and magma. It deals with the formation and classification of volcanoes and also deals with the causes of these eruptions and the effect of the eruptions on other large-scale geologic processes. This chapter will provide an integrated understanding of volcanology.

Volcanology (also spelled vulcanology) is the study of volcanoes, lava, magma, and related geological, geophysical and geochemical phenomena. The term *volcanology* is derived from the Latin word *vulcan*. Vulcan was the ancient Roman god of fire.

A volcanologist sampling lava using a rock hammer and a bucket of water

A volcanologist is a geologist who studies the eruptive activity and formation of volcanoes, and their current and historic eruptions. Volcanologists frequently visit volcanoes, especially active ones, to observe volcanic eruptions, collect eruptive products including tephra (such as ash or pumice), rock and lava samples. One major focus of enquiry is the prediction of eruptions; there is currently no accurate way to do this, but predicting eruptions, like predicting earthquakes, could save many lives.

Modern Volcanology

In 1841, the first volcanological observatory, the Vesuvius Observatory, was founded in the Kingdom of the Two Sicilies.

Seismic observations are made using seismographs deployed near volcanic areas, watching out for increased seismicity during volcanic events, in particular looking for long period harmonic tremors, which signal magma movement through volcanic conduits.

Volcanologist examining tephra horizons in south-central Iceland.

Surface deformation monitoring includes the use of geodetic techniques such as leveling, tilt, strain, angle and distance measurements through tiltmeters, total stations and EDMs. This also includes GNSS observations and InSAR. Surface deformation indicates magma upwelling: increased magma supply produces bulges in the volcanic center's surface.

Gas emissions may be monitored with equipment including portable ultra-violet spectrometers (COSPEC, now superseded by the miniDOAS), which analyzes the presence of volcanic gases such as sulfur dioxide; or by infra-red spectroscopy (FTIR). Increased gas emissions, and more particularly changes in gas compositions, may signal an impending volcanic eruption.

Temperature changes are monitored using thermometers and observing changes in thermal properties of volcanic lakes and vents, which may indicate upcoming activity.

Satellites are widely used to monitor volcanoes, as they allow a large area to be monitored easily. They can measure the spread of an ash plume, such as the one from Eyjafjallajokull's 2010 eruption, as well as SO_2 emissions. InSAR and thermal imaging can monitor large, scarcely populated areas where it would be too expensive to maintain instruments on the ground.

Other geophysical techniques (electrical, gravity and magnetic observations) include monitoring fluctuations and sudden change in resistivity, gravity anomalies or magnetic anomaly patterns that may indicate volcano-induced faulting and magma upwelling.

Stratigraphic analyses includes analyzing tephra and lava deposits and dating these to give volcano eruption patterns, with estimated cycles of intense activity and size of eruptions.

History

Volcanology has an extensive history. The earliest known recording of a volcanic eruption may be on a wall painting dated to about 7,000 BCE found at the Neolithic site at

Çatal Höyük in Anatolia, Turkey. This painting has been interpreted as a depiction of an erupting volcano, with a cluster of houses below shows a twin peaked volcano in eruption, with a town at its base (though archaeologists now question this interpretation). The volcano may be either Hasan Dağ, or its smaller neighbour, Melendiz Dağ.

Greco-Roman Philosophy

The classical world of Greece and the early Roman Empire explained volcanoes as sites of various gods. Greeks considered that Hephaestus, the god of fire, sat below the volcano Etna, forging the weapons of Zeus. The Greek word used to describe volcanoes was *etna*, or *hiera*, after Heracles, the son of Zeus. The Roman poet Virgil, in interpreting the Greek mythos, held that the giant Enceladus was buried beneath Etna by the goddess Athena as punishment for rebellion against the gods; the mountain's rumblings were his tormented cries, the flames his breath and the tremors his railing against the bars of his prison. Enceladus' brother Mimas was buried beneath Vesuvius by Hephaestus, and the blood of other defeated giants welled up in the Phlegrean Fields surrounding Vesuvius.

Eruption of Vesuvius in 1822. The eruption of CE 79 would have appeared very similar.

The Greek philosopher Empedocles (c. 490-430 BCE) saw the world divided into four elemental forces, of Earth, Air, Fire and Water. Volcanoes, Empedocles maintained, were the manifestation of Elemental Fire. Plato contended that channels of hot and cold waters flow in inexhaustible quantities through subterranean rivers. In the depths of the earth snakes a vast river of fire, the *Pyriphlegethon*, which feeds all the world's volcanoes. Aristotle considered underground fire as the result of "the...friction of the wind when it plunges into narrow passages."

Wind played a key role in volcano explanations until the 16th century. Lucretius, a Roman philosopher, claimed Etna was completely hollow and the fires of the underground driven by a fierce wind circulating near sea level. Ovid believed that the flame was fed from "fatty foods" and eruptions stopped when the food ran out. Vitruvius contended that sulfur, alum and bitumen fed the deep fires. Observations by Pliny the Elder noted the presence of earthquakes preceded an eruption; he died in the eruption of Vesuvius in 79 CE while investigating it at Stabiae. His nephew, Pliny the Younger gave detailed descriptions of the eruption in which his uncle died, attributing his death to the effects of toxic gases. Such eruptions have been named Plinian in honour of the two authors.

Renaissance Observations

Nuées ardentes were described from the Azores in 1580. Georgius Agricola argued the rays of the sun, as later proposed by Descartes had nothing to do with volcanoes. Agricola believed vapor under pressure caused eruptions of 'mointain oil' and basalt.

Jesuit Athanasius Kircher (1602–1680) witnessed eruptions of Mount Etna and Stromboli, then visited the crater of Vesuvius and published his view of an Earth with a central fire connected to numerous others caused by the burning of sulfur, bitumen and coal.

Johannes Kepler considered volcanoes as conduits for the tears and excrement of the Earth, voiding bitumen, tar and sulfur. Descartes, pronouncing that God had created the Earth in an instant, declared he had done so in three layers; the fiery depths, a layer of water, and the air. Volcanoes, he said, were formed where the rays of the sun pierced the earth.

Science wrestled with the ideas of the combustion of pyrite with water, that rock was solidified bitumen, and with notions of rock being formed from water (Neptunism). Of the volcanoes then known, all were near the water, hence the action of the sea upon the land was used to explain volcanism.

Interaction with Religion and Mythology

Pele's hair caught on a radio antenna mounted on the south rim, Hawaii, July 22, 2005

Tribal legends of volcanoes abound from the Pacific Ring of Fire and the Americas, usually invoking the forces of the supernatural or the divine to explain the violent outbursts of volcanoes. Taranaki and Tongariro, according to Māori mythology, were lovers who fell in love with Pihanga, and a spiteful jealous fight ensued. Māori will not to this day live between Tongariro and Taranaki for fear of the dispute flaring up again.

In the Hawaiian religion, Pele is the goddess of volcanoes and a popular figure in Hawaiian mythology. Pele was used for various scientific terms as for Pele's hair, Pele's tears, and Limu o Pele (Pele's seaweed). A volcano on the Jovian moon Io is also named Pele.

Saint Agatha is patron saint of Catania, close to mount Etna, and an important highly venerated (till today) example of virgin martyrs of Christian antiquity. 253 CE, one year after her violent death, the stilling of an eruption of Mt. Etna was attributed to her intercession. Catania was however nearly completely destroyed by the eruption of Mt. Etna in 1169, and over 15,000 of its inhabitants died. Nevertheless, she was invoked again 1669 and, for an outbreak danginering Nicolosi in 1886. The way she is invoked and dealt with in Italian Folk religion, a sort of quid pro quo way approach to saints, has been related (in the tradition of James Frazer) to earlier pagan believes.

In 1660 the eruption of Vesuvius rained twinned pyroxene crystals and ash upon the nearby villages. The crystals resembled the crucifix and this was interpreted as the work of Saint Januarius. In Naples, the relics of St Januarius are paraded through town at every major eruption of Vesuvius. The register of these processions and the 1779 and 1794 diary of Father Antonio Piaggio allowed British diplomat and amateur naturalist Sir William Hamilton to provide a detailed chronology and description of Vesuvius' eruptions.

References

- Peter Francis and Clive Oppenheimer, Volcanoes, Oxford University Press, USA 2003, 2nd ed., ISBN 0-19-925469-9
- Foley O.F.M., Leonard. Saint of the Day, (revised by Pat McCloskey O.F.M.), Franciscan Media, ISBN 978-0-86716-887-7
- H. Arlo Nimmo (2011). Pele, Volcano Goddess of Hawai'i: A History. McFarland. p. 208. ISBN 0-7864-6347-3.
- Kirsch, Johann Peter. "St. Agatha." The Catholic Encyclopedia. Vol. 1. New York: Robert Appleton Company, 1907. 25 April 2013

2

Volcano: An Integrated Study

Volcanoes are a break in the crust of the Earth that allows lava, volcanic ash, and magma to escape from a magma chamber underneath the surface of earth. These eruptions do not only occur on the Earth but on every planetary mass object. The chapter on volcanoes offers an insightful focus, keeping in mind the theme of the chapter.

Volcano

A volcano is a rupture in the crust of a planetary-mass object, such as Earth, that allows hot lava, volcanic ash, and gases to escape from a magma chamber below the surface.

Cleveland Volcano in the Aleutian Islands of Alaska photographed from the International Space Station, May 2006

Earth's volcanoes occur because its crust is broken into 17 major, rigid tectonic plates that float on a hotter, softer layer in its mantle. Therefore, on Earth, volcanoes are generally found where tectonic plates are diverging or converging. For example, a mid-oceanic ridge, such as the Mid-Atlantic Ridge, has volcanoes caused by divergent tectonic plates pulling apart; the Pacific Ring of Fire has volcanoes caused by convergent tectonic plates coming together. Volcanoes can also form where there is stretching and thinning of the crust's interior plates, e.g., in the East African Rift and the Wells Gray-Clearwater volcanic field and Rio Grande Rift in North America. This type of volcanism falls under the umbrella of "plate hypothesis" volcanism. Volcanism away from plate boundaries has also been explained as mantle plumes. These so-called "hotspots", for example Hawaii, are postulated to arise from upwelling diapirs with magma from the core–mantle boundary, 3,000 km deep in the Earth. Volcanoes are usually not created where two tectonic plates slide past one another.

Ash plumes reached a height of 19 kilometres (12 mi) during the climactic explosive eruption at Mount Pinatubo, Philippines in 1991.

A 2007 eruptive column at Mount Etna producing volcanic ash, pumice and lava bombs

Ubinas Volcano

Santa Ana Volcano, El Salvador. A close-up aerial view of the nested summit calderas and craters, along with the crater lake.

Erupting volcanoes can pose many hazards, not only in the immediate vicinity of the eruption. One such hazard is that volcanic ash can be a threat to aircraft, in particular those with jet engines where ash particles can be melted by the high operating temperature; the melted particles then adhere to the turbine blades and alter their shape, disrupting the operation of the turbine. Large eruptions can affect temperature as ash and droplets of sulfuric acid obscure the sun and cool the Earth's lower atmosphere (or troposphere); however, they also absorb heat radiated up from the Earth, thereby warming the upper atmosphere (or stratosphere). Historically, so-called volcanic winters have caused catastrophic famines.

Etymology

The word *volcano* is derived from the name of Vulcano, a volcanic island in the Aeolian Islands of Italy whose name in turn comes from Vulcan, the god of fire in Roman mythology. The study of volcanoes is called volcanology, sometimes spelled *vulcanology*.

Plate Tectonics

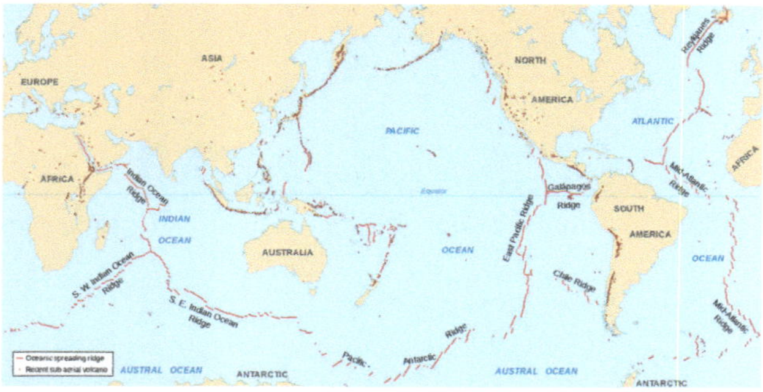

Map showing the divergent plate boundaries (OSR – Oceanic Spreading Ridges) and recent sub-aerial volcanoes

Divergent Plate Boundaries

At the mid-oceanic ridges, two tectonic plates diverge from one another as new oceanic crust is formed by the cooling and solidifying of hot molten rock. Because the crust is very thin at these ridges due to the pull of the tectonic plates, the release of pressure leads to adiabatic expansion and the partial melting of the mantle, causing volcanism and creating new oceanic crust. Most divergent plate boundaries are at the bottom of the oceans; therefore, most volcanic activity is submarine, forming new seafloor. Black smokers (also known as deep sea vents) are evidence of this kind of volcanic activity. Where the mid-oceanic ridge is above sea-level, volcanic islands are formed, for example, Iceland.

Convergent Plate Boundaries

Subduction zones are places where two plates, usually an oceanic plate and a continental plate, collide. In this case, the oceanic plate subducts, or submerges under the continental plate forming a deep ocean trench just offshore. In a process called flux melting, water released from the subducting plate lowers the melting temperature of the overlying mantle wedge, creating magma. This magma tends to be very viscous due to its high silica content, so it often does not reach the surface but cools at depth. When it does reach the surface, a volcano is formed. Typical examples of this kind of volcano are Mount Etna and the volcanoes in the Pacific Ring of Fire.

"Hotspots"

"Hotspots" is the name given to volcanic areas believed to be formed by mantle plumes, which are hypothesized to be columns of hot material rising from the core-mantle boundary in a fixed space that causes large-volume melting. Because tectonic plates move across them, each volcano becomes dormant and is eventually reformed as the plate advances over the postulated plume. The Hawaiian Islands have been suggested to have been formed in such a manner, as well as the Snake River Plain, with the Yellowstone Caldera being the part of the North American plate currently above the hot spot. This theory is currently under criticism, however.

Volcanic Features

Lakagigar fissure vent in Iceland, source of the major world climate alteration of 1783–84

Skjaldbreiður, a shield volcano whose name means "broad shield"

The most common perception of a volcano is of a conical mountain, spewing lava and poisonous gases from a crater at its summit; however, this describes just one of the many types of volcano. The features of volcanoes are much more complicated and their structure and behavior depends on a number of factors. Some volcanoes have rugged peaks formed by lava domes rather than a summit crater while others have landscape features such as massive plateaus. Vents that issue volcanic material (including lava and ash) and gases (mainly steam and magmatic gases) can develop anywhere on the landform and may give rise to smaller cones such as Pu'u'Ō'ō on a flank of Hawaii's Kīlauea. Other types of volcano include cryovolcanoes (or ice volcanoes), particularly on some moons of Jupiter, Saturn, and Neptune; and mud volcanoes, which are formations often not associated with known magmatic activity. Active mud volcanoes tend to involve temperatures much lower than those of igneous volcanoes except when the mud volcano is actually a vent of an igneous volcano.

Fissure Vents

Volcanic fissure vents are flat, linear fractures through which lava emerges.

Shield Volcanoes

Shield volcanoes, so named for their broad, shield-like profiles, are formed by the eruption of low-viscosity lava that can flow a great distance from a vent. They generally do not explode catastrophically. Since low-viscosity magma is typically low in silica, shield volcanoes are more common in oceanic than continental settings. The Hawaiian volcanic chain is a series of shield cones, and they are common in Iceland, as well.

Lava Domes

Lava domes are built by slow eruptions of highly viscous lava. They are sometimes formed within the crater of a previous volcanic eruption, as in the case of Mount Saint Helens, but can also form independently, as in the case of Lassen Peak. Like stratovolcanoes, they can produce violent, explosive eruptions, but their lava generally does not flow far from the originating vent.

Cryptodomes

Cryptodomes are formed when viscous lava is forced upward causing the surface to bulge. The 1980 eruption of Mount St. Helens was an example; lava beneath the surface of the mountain created an upward bulge which slid down the north side of the mountain.

Volcanic Cones (Cinder Cones)

Volcanic cones or cinder cones result from eruptions of mostly small pieces of scoria and pyroclastics (both resemble cinders, hence the name of this volcano type) that build up around the vent. These can be relatively short-lived eruptions that produce a cone-shaped hill perhaps 30 to 400 meters high. Most cinder cones erupt only once. Cinder cones may form as flank vents on larger volcanoes, or occur on their own. Parícutin in Mexico and Sunset Crater in Arizona are examples of cinder cones. In New Mexico, Caja del Rio is a volcanic field of over 60 cinder cones.

Izalco (volcano), located in the Cordillera de Apaneca volcanic range complex in El Salvador. Only a few generations old, Izalco is the youngest and best known cone volcano. Izalco erupted almost continuously from 1770 (when it formed) to 1958, earning it the nickname of "Lighthouse of the Pacific".

Based on satellite images it was suggested that cinder cones might occur on other terrestrial bodies in the Solar system too; on the surface of Mars and the Moon.

Stratovolcanoes (Composite Volcanoes)

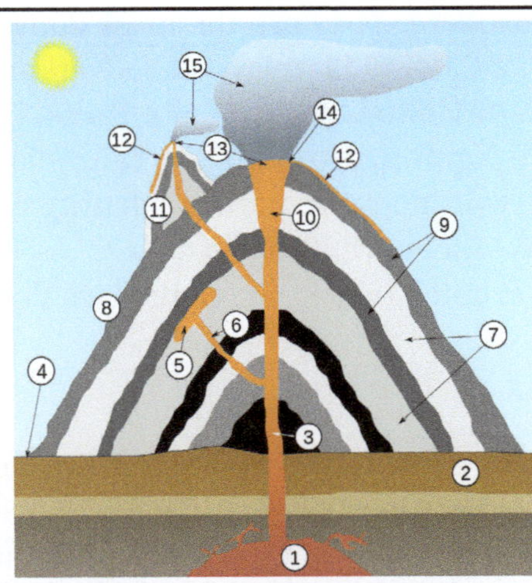

Cross-section through a stratovolcano (vertical scale is exaggerated):

1. Large magma chamber 2. Bedrock 3. Conduit (pipe) 4. Base 5. Sill	6. Dike 7. Layers of ash emitted by the volcano 8. Flank 9. Layers of lava emitted by the volcano 10. Throat	11. Parasitic cone 12. Lava flow 13. Vent 14. Crater 15. Ash cloud

Stratovolcanoes or composite volcanoes are tall conical mountains composed of lava flows and other ejecta in alternate layers, the strata that gives rise to the name. Stratovolcanoes are also known as composite volcanoes because they are created from multiple structures during different kinds of eruptions. Strato/composite volcanoes are made of cinders, ash, and lava. Cinders and ash pile on top of each other, lava flows on top of the ash, where it cools and hardens, and then the process repeats. Classic examples include Mount Fuji in Japan, Mayon Volcano in the Philippines, and Mount Vesuvius and Stromboli in Italy.

Throughout recorded history, ash produced by the explosive eruption of stratovolcanoes has posed the greatest volcanic hazard to civilizations. Not only do stratovolcanoes have greater pressure build up from the underlying lava flow than shield volcanoes, but their fissure vents and monogenetic volcanic fields (volcanic cones) have more powerful eruptions, as they are many times under extension. They are also steeper than shield volcanoes, with slopes of 30–35° compared to slopes of generally 5–10°, and their loose tephra are material for dangerous lahars. Large pieces of tephra are called volcanic bombs. Big bombs can measure more than 4 feet(1.2 meters) across and weigh several tons.

Supervolcanoes

A supervolcano usually has a large caldera and can produce devastation on an enormous, sometimes continental, scale. Such volcanoes are able to severely cool global temperatures for many years after the eruption due to the huge volumes of sulfur and ash released into the atmosphere. They are the most dangerous type of volcano. Examples include: Yellowstone Caldera in Yellowstone National Park and Valles Caldera in New Mexico (both western United States); Lake Taupo in New Zealand; Lake Toba in Sumatra, Indonesia; and Ngorongoro Crater in Tanzania. Because of the enormous area they may cover, supervolcanoes are hard to identify centuries after an eruption. Similarly, large igneous provinces are also considered supervolcanoes because of the vast amount of basalt lava erupted (even though the lava flow is non-explosive).

Submarine Volcanoes

Submarine volcanoes are common features of the ocean floor. In shallow water, active volcanoes disclose their presence by blasting steam and rocky debris high above the ocean's surface. In the ocean's deep, the tremendous weight of the water above prevents the explosive release of steam and gases; however, they can be detected by hydrophones and discoloration of water because of volcanic gases. Pillow lava is a common eruptive product of submarine volcanoes and is characterized by thick sequences of discontinuous pillow-shaped masses which form under water. Even large submarine eruptions may not disturb the ocean surface due to the rapid cooling effect and increased buoyancy of water (as compared to air) which often causes volcanic vents to form steep pillars on the ocean floor. Hydrothermal vents are common near these volcanoes, and some support peculiar ecosystems based on dissolved minerals. Over time, the formations created by submarine volcanoes may become so large that they break the ocean surface as new islands or floating pumice rafts.

Subglacial Volcanoes

Subglacial volcanoes develop underneath icecaps. They are made up of flat lava which flows at the top of extensive pillow lavas and palagonite. When the icecap melts, the lava on top collapses, leaving a flat-topped mountain. These volcanoes are also called table mountains, tuyas, or (uncommonly) mobergs. Very good examples of this type of volcano can be seen in Iceland, however, there are also tuyas in British Columbia. The origin of the term comes from Tuya Butte, which is one of the several tuyas in the area of the Tuya River and Tuya Range in northern British Columbia. Tuya Butte was the first such landform analyzed and so its name has entered the geological literature for this kind of volcanic formation. The Tuya Mountains Provincial Park was recently established to protect this unusual landscape, which lies north of Tuya Lake and south of the Jennings River near the boundary with the Yukon Territory.

Mud Volcanoes

Mud volcanoes or mud domes are formations created by geo-excreted liquids and gases, although there are several processes which may cause such activity. The largest structures are 10 kilometers in diameter and reach 700 meters high.

Erupted Material

Pāhoehoe lava flow on Hawaii. The picture shows overflows of a main lava channel.

The Stromboli stratovolcano off the coast of Sicily has erupted continuously for thousands of years, giving rise to the term strombolian eruption.

San Miguel (volcano), El Salvador. On December 29, 2013, San Miguel volcano, also known as "Chaparrastique", erupted at 10:30 local time, spewing a large column of ash and smoke into the sky; the eruption, the first in 11 years, was seen from space and prompted the evacuation of thousands of people living in a 3 km radius around the volcano.

Ash plume from San Miguel (volcano) "Chaparrastique", seen from a satellite, as it heads towards the Pacific Ocean from the El Salvador Central America coast, December 29, 2013

Lava Composition

Another way of classifying volcanoes is by the *composition of material erupted* (lava), since this affects the shape of the volcano. Lava can be broadly classified into 4 different compositions (Cas & Wright, 1987):

- If the erupted magma contains a high percentage (>63%) of silica, the lava is called felsic.
    - Felsic lavas (dacites or rhyolites) tend to be highly viscous (not very fluid) and are erupted as domes or short, stubby flows. Viscous lavas tend to form stratovolcanoes or lava domes. Lassen Peak in California is an example of a volcano formed from felsic lava and is actually a large lava dome.
    - Because siliceous magmas are so viscous, they tend to trap volatiles (gases) that are present, which cause the magma to erupt catastrophically, eventually forming stratovolcanoes. Pyroclastic flows (ignimbrites) are highly hazardous products of such volcanoes, since they are composed of molten volcanic ash too heavy to go up into the atmosphere, so they hug the volcano's slopes and travel far from their vents during large eruptions. Temperatures as high as 1,200 °C are known to occur in py-

roclastic flows, which will incinerate everything flammable in their path and thick layers of hot pyroclastic flow deposits can be laid down, often up to many meters thick. Alaska's Valley of Ten Thousand Smokes, formed by the eruption of Novarupta near Katmai in 1912, is an example of a thick pyroclastic flow or ignimbrite deposit. Volcanic ash that is light enough to be erupted high into the Earth's atmosphere may travel many kilometres before it falls back to ground as a tuff.

- If the erupted magma contains 52–63% silica, the lava is of *intermediate* composition.
    - These "andesitic" volcanoes generally only occur above subduction zones (e.g. Mount Merapi in Indonesia).
    - Andesitic lava is typically formed at convergent boundary margins of tectonic plates, by several processes:
        - Hydration melting of peridotite and fractional crystallization
        - Melting of subducted slab containing sediments
        - Magma mixing between felsic rhyolitic and mafic basaltic magmas in an intermediate reservoir prior to emplacement or lava flow.

Sarychev Peak eruption, Matua Island, oblique satellite view

- If the erupted magma contains <52% and >45% silica, the lava is called mafic (because it contains higher percentages of magnesium (Mg) and iron (Fe)) or basaltic. These lavas are usually much less viscous than rhyolitic lavas, depending on their eruption temperature; they also tend to be hotter than felsic lavas. Mafic lavas occur in a wide range of settings:
    - At mid-ocean ridges, where two oceanic plates are pulling apart, basaltic lava erupts as pillows to fill the gap;
    - Shield volcanoes (e.g. the Hawaiian Islands, including Mauna Loa and Kilauea), on both oceanic and continental crust;

 - As continental flood basalts.

- Some erupted magmas contain <=45% silica and produce ultramafic lava. Ultramafic flows, also known as komatiites, are very rare; indeed, very few have been erupted at the Earth's surface since the Proterozoic, when the planet's heat flow was higher. They are (or were) the hottest lavas, and probably more fluid than common mafic lavas.

Lava Texture

Two types of lava are named according to the surface texture: lava and pāhoehoe, both Hawaiian words. lava is character-ized by a rough, clinkery surface and is the typical texture of viscous lava flows. Howev-er, even basaltic or mafic flows can be erupted as lava flows, particularly if the erup-tion rate is high and the slope is steep.

Pāhoehoe is characterized by its smooth and often ropey or wrinkly surface and is generally formed from more fluid lava flows. Usually, only mafic flows will erupt as pāhoehoe, since they often erupt at higher temperatures or have the proper chemical makeup to allow them to flow with greater fluidity.

Volcanic Activity

Fresco with Mount Vesuvius behind Bacchus and Agathodaemon,
as seen in Pompeii's House of the Centenary

Popular Classification of Volcanoes

A popular way of classifying magmatic volcanoes is by their frequency of eruption, with those that erupt regularly called active, those that have erupted in historical times but are now quiet called dormant or inactive, and those that have not erupted in historical times called extinct. However, these popular classifications—extinct in particular—are practically meaningless to scientists. They use classifications which refer to a particular volcano's formative and eruptive processes and resulting shapes, which was explained above.

Active

There is no consensus among volcanologists on how to define an "active" volcano. The lifespan of a volcano can vary from months to several million years, making such a distinction sometimes meaningless when compared to the lifespans of humans or even civilizations. For example, many of Earth's volcanoes have erupted dozens of times in the past few thousand years but are not currently showing signs of eruption. Given the long lifespan of such volcanoes, they are very active. By human lifespans, however, they are not.

Scientists usually consider a volcano to be *erupting* or *likely to erupt* if it is currently erupting, or showing signs of unrest such as unusual earthquake activity or significant new gas emissions. Most scientists consider a volcano *active* if it has erupted in the last 10,000 years (Holocene times) – the Smithsonian Global Volcanism Program uses this definition of *active*. Most volcanoes are situated on the Pacific Ring of Fire. An estimated 500 million people live near active volcanoes.

Historical time (or recorded history) is another timeframe for *active*. The *Catalogue of the Active Volcanoes of the World*, published by the International Association of Volcanology, uses this definition, by which there are more than 500 active volcanoes. However, the span of recorded history differs from region to region. In China and the Mediterranean, it reaches back nearly 3,000 years, but in the Pacific Northwest of the United States and Canada, it reaches back less than 300 years, and in Hawaii and New Zealand, only around 200 years.

Kīlauea lava entering the sea.

Lava flows at Holuhraun, Iceland, September 2014

As of 2013, the following are considered Earth's most active volcanoes:

- Kīlauea, the famous Hawaiian volcano, has been in continuous, effusive eruption since 1983, and has the longest-observed lava lake.
- Mount Etna and nearby Stromboli, two Mediterranean volcanoes in "almost continuous eruption" since antiquity.
- Mount Yasur, in Vanuatu, has been erupting "nearly continuously" for over 800 years.

The longest currently ongoing (but not necessarily continuous) volcanic eruptive phases are:

- Mount Yasur, 111 years
- Mount Etna, 109 years
- Stromboli, 108 years
- Santa María, 101 years
- Sangay, 94 years

Other very active volcanoes include:

- Mount Nyiragongo and its neighbor, Nyamuragira, are Africa's most active volcanoes

Nyiragongo's lava lake

- Piton de la Fournaise, in Réunion, erupts frequently enough to be a tourist attraction.
- Erta Ale, in the Afar Triangle, has maintained a lava lake since at least 1906.
- Mount Erebus, in Antarctica, has maintained a lava lake since at least 1972.
- Mount Merapi
- Whakaari / White Island, has been in continuous state of smoking since its discovery in 1769.

- Ol Doinyo Lengai
- Ambrym
- Arenal Volcano
- Pacaya
- Klyuchevskaya Sopka
- Sheveluch

Extinct

Extinct volcanoes are those that scientists consider unlikely to erupt again because the volcano no longer has a magma supply. Examples of extinct volcanoes are many volcanoes on the Hawaiian – Emperor seamount chain in the Pacific Ocean, Hohentwiel, Shiprock and the Zuidwal volcano in the Netherlands. Edinburgh Castle in Scotland is famously located atop an extinct volcano. Otherwise, whether a volcano is truly extinct is often difficult to determine. Since "supervolcano" calderas can have eruptive lifespans sometimes measured in millions of years, a caldera that has not produced an eruption in tens of thousands of years is likely to be considered dormant instead of extinct. Some volcanologists refer to extinct volcanoes as inactive, though the term is now more commonly used for dormant volcanoes once thought to be extinct.

Fourpeaked volcano, Alaska, in September 2006 after being thought extinct for over 10,000 years

Mount Rinjani eruption in 1994, in Lombok, Indonesia

Dormant

It is difficult to distinguish an extinct volcano from a dormant (inactive) one. Volcanoes are often considered to be extinct if there are no written records of its activity. Nevertheless, volcanoes may remain dormant for a long period of time. For example, Yellowstone has a repose/recharge period of around 700,000 years, and Toba of around 380,000 years. Vesuvius was described by Roman writers as having been covered with gardens and vineyards before its eruption of AD 79, which destroyed the towns of Herculaneum and Pompeii. Before its catastrophic eruption of 1991, Pinatubo was an inconspicuous volcano, unknown to most people in the surrounding areas. Two other examples are the long-dormant Soufrière Hills volcano on the island of Montserrat, thought to be extinct before activity resumed in 1995 and Fourpeaked Mountain in Alaska, which, before its September 2006 eruption, had not erupted since before 8000 BC and had long been thought to be extinct.

Narcondam Island, India, is classified as a dormant volcano by the Geological Survey of India

Technical Classification of Volcanoes

Volcanic-alert Level

The three common popular classifications of volcanoes can be subjective and some volcanoes thought to have been extinct have erupted again. To help prevent people from falsely believing they are not at risk when living on or near a volcano, countries have adopted new classifications to describe the various levels and stages of volcanic activity. Some alert systems use different numbers or colors to designate the different stages. Other systems use colors and words. Some systems use a combination of both.

Volcano Warning Schemes of the United States

The United States Geological Survey (USGS) has adopted a common system nationwide for characterizing the level of unrest and eruptive activity at volcanoes. The new volcano alert-level system classifies volcanoes now as being in a normal, advisory, watch or warning stage. Additionally, colors are used to denote the amount of ash produced. Details of the U.S. system can be found at Volcano warning schemes of the United States.

Decade Volcanoes

Koryaksky volcano towering over Petropavlovsk-Kamchatsky on Kamchatka Peninsula, Far Eastern Russia

The Decade Volcanoes are 17 volcanoes identified by the International Association of Volcanology and Chemistry of the Earth's Interior (IAVCEI) as being worthy of particular study in light of their history of large, destructive eruptions and proximity to populated areas. They are named Decade Volcanoes because the project was initiated as part of the United Nations-sponsored International Decade for Natural Disaster Reduction. The 17 current Decade Volcanoes are

- Avachinsky-Koryaksky (grouped together), Kamchatka, Russia
- Nevado de Colima, Jalisco and Colima, Mexico
- Mount Etna, Sicily, Italy
- Galeras, Nariño, Colombia
- Mauna Loa, Hawaii, USA
- Mount Merapi, Central Java, Indonesia
- Mount Nyiragongo, Democratic Republic of the Congo
- Mount Rainier, Washington, USA
- Sakurajima, Kagoshima Prefecture, Japan
- Santa Maria/Santiaguito, Guatemala
- Santorini, Cyclades, Greece
- Taal Volcano, Luzon, Philippines
- Teide, Canary Islands, Spain
- Ulawun, New Britain, Papua New Guinea
- Mount Unzen, Nagasaki Prefecture, Japan
- Vesuvius, Naples, Italy

The Deep Earth Carbon Degassing Project, an initiative of the Deep Carbon Observatory, monitors 9 volcanoes, 2 of which are Decade volcanoes. The focus of the Deep Earth Carbon Degassing Project is to use Multi-Component Gas Analyzer System instruments to measure CO_2/SO_2 ratios in real-time and in high-resolution to allow detection of the pre-eruptive degassing of rising magmas, improving prediction of volcanic activity.

Effects of Volcanoes

There are many different types of volcanic eruptions and associated activity: phreatic eruptions (steam-generated eruptions), explosive eruption of high-silica lava (e.g., rhyolite), effusive eruption of low-silica lava (e.g., basalt), pyroclastic flows, lahars (debris flow) and carbon dioxide emission. All of these activities can pose a hazard to humans.

Earthquakes, hot springs, fumaroles, mud pots and geysers often accompany volcanic activity.

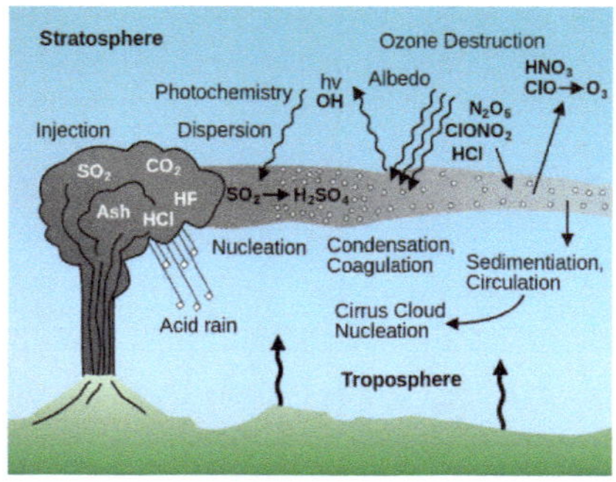

Schematic of volcano injection of aerosols and gases

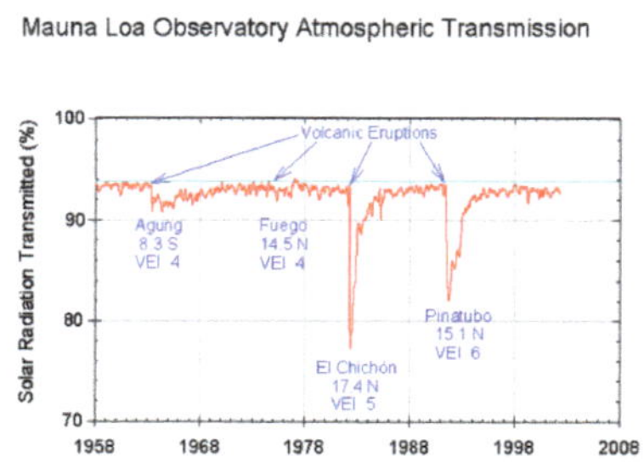

Solar radiation graph 1958–2008, showing how the radiation is reduced after major volcanic eruptions

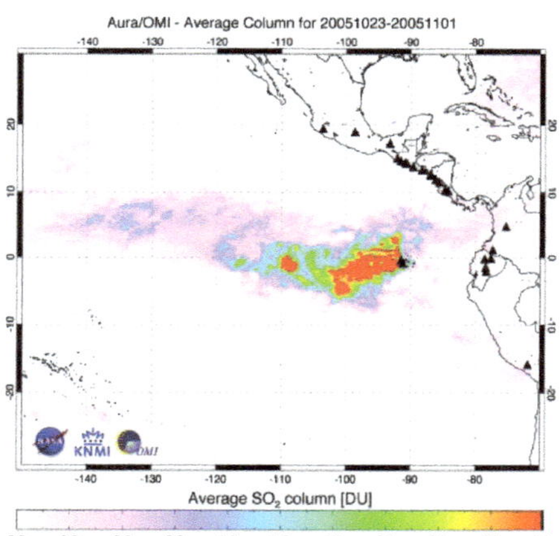

Sulfur dioxide concentration over the Sierra Negra Volcano, Galapagos Islands, during an eruption in October 2005

Volcanic Gases

The concentrations of different volcanic gases can vary considerably from one volcano to the next. Water vapor is typically the most abundant volcanic gas, followed by carbon dioxide and sulfur dioxide. Other principal volcanic gases include hydrogen sulfide, hydrogen chloride, and hydrogen fluoride. A large number of minor and trace gases are also found in volcanic emissions, for example hydrogen, carbon monoxide, halocarbons, organic compounds, and volatile metal chlorides.

Large, explosive volcanic eruptions inject water vapor (H_2O), carbon dioxide (CO_2), sulfur dioxide (SO_2), hydrogen chloride (HCl), hydrogen fluoride (HF) and ash (pulverized rock and pumice) into the stratosphere to heights of 16–32 kilometres (10–20 mi) above the Earth's surface. The most significant impacts from these injections come from the conversion of sulfur dioxide to sulfuric acid (H_2SO_4), which condenses rapidly in the stratosphere to form fine sulfate aerosols. It is worth mentioning that the SO_2 emissions alone of two different eruptions are sufficient to compare their potential climatic impact. The aerosols increase the Earth's albedo—its reflection of radiation from the Sun back into space – and thus cool the Earth's lower atmosphere or troposphere; however, they also absorb heat radiated up from the Earth, thereby warming the stratosphere. Several eruptions during the past century have caused a decline in the average temperature at the Earth's surface of up to half a degree (Fahrenheit scale) for periods of one to three years – sulfur dioxide from the eruption of Huaynaputina probably caused the Russian famine of 1601–1603.

Significant Consequences

One proposed volcanic winter happened c. 70,000 years ago following the supereruption of Lake Toba on Sumatra island in Indonesia. According to the Toba catastrophe theory to which some anthropologists and archeologists subscribe, it had global consequences, killing most humans then alive and creating a population bottleneck that affected the genetic inheritance of all humans today. The 1815 eruption of Mount Tambora created global climate anomalies that became known as the "Year Without a Summer" because of the effect on North American and European weather. Agricultural crops failed and livestock died in much of the Northern Hemisphere, resulting in one of the worst famines of the 19th century. The freezing winter of 1740–41, which led to widespread famine in northern Europe, may also owe its origins to a volcanic eruption.

It has been suggested that volcanic activity caused or contributed to the End-Ordovician, Permian-Triassic, Late Devonian mass extinctions, and possibly others. The massive eruptive event which formed the Siberian Traps, one of the largest known volcanic events of the last 500 million years of Earth's geological history, continued for a million years and is considered to be the likely cause of the "Great Dying" about 250 million years ago, which is estimated to have killed 90% of species existing at the time.

Acid Rain

The sulfate aerosols also promote complex chemical reactions on their surfaces that alter chlorine and nitrogen chemical species in the stratosphere. This effect, together with increased stratospheric chlorine levels from chlorofluorocarbon pollution, generates chlorine monoxide (ClO), which destroys ozone (O_3). As the aerosols grow and coagulate, they settle down into the upper troposphere where they serve as nuclei for cirrus clouds and further modify the Earth's radiation balance. Most of the hydrogen chloride (HCl) and hydrogen fluoride (HF) are dissolved in water droplets in the eruption cloud and quickly fall to the ground as acid rain. The injected ash also falls rapidly from the stratosphere; most of it is removed within several days to a few weeks. Finally, explosive volcanic eruptions release the greenhouse gas carbon dioxide and thus provide a deep source of carbon for biogeochemical cycles.

Ash plume rising from Eyjafjallajökull on April 17, 2010

Gas emissions from volcanoes are a natural contributor to acid rain. Volcanic activity releases about 130 to 230 teragrams (145 million to 255 million short tons) of carbon dioxide each year. Volcanic eruptions may inject aerosols into the Earth's atmosphere. Large injections may cause visual effects such as unusually colorful sunsets and affect global climate mainly by cooling it. Volcanic eruptions also provide the benefit of adding nutrients to soil through the weathering process of volcanic rocks. These fertile soils assist the growth of plants and various crops. Volcanic eruptions can also create new islands, as the magma cools and solidifies upon contact with the water.

Hazards

Ash thrown into the air by eruptions can present a hazard to aircraft, especially jet aircraft where the particles can be melted by the high operating temperature; the melted particles then adhere to the turbine blades and alter their shape, disrupting the operation of the turbine. Dangerous encounters in 1982 after the eruption of Galunggung in Indonesia, and 1989 after the eruption of Mount Redoubt in Alaska raised awareness of this phenomenon. Nine Volcanic Ash Advisory Centers were established by the International Civil Aviation Organization to monitor ash clouds and advise pilots accordingly. The 2010 eruptions of Eyjafjallajökull caused major disruptions to air travel in Europe.

Volcanoes on other Celestial Bodies

The Earth's Moon has no large volcanoes and no current volcanic activity, although recent evidence suggests it may still possess a partially molten core. However, the Moon does have many volcanic features such as maria (the darker patches seen on the moon), rilles and domes.

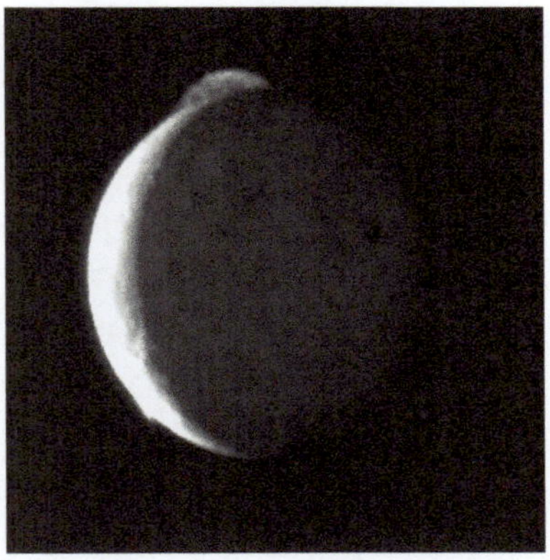

The Tvashtar volcano erupts a plume 330 km (205 mi) above the surface of Jupiter's moon Io.

The planet Venus has a surface that is 90% basalt, indicating that volcanism played a major role in shaping its surface. The planet may have had a major global resurfacing event about 500 million years ago, from what scientists can tell from the density of impact craters on the surface. Lava flows are widespread and forms of volcanism not present on Earth occur as well. Changes in the planet's atmosphere and observations of lightning have been attributed to ongoing volcanic eruptions, although there is no confirmation of whether or not Venus is still volcanically active. However, radar sounding by the Magellan probe revealed evidence for comparatively recent volcanic activity at Venus's highest volcano Maat Mons, in the form of ash flows near the summit and on the northern flank.

Olympus Mons (Latin, "Mount Olympus"), located on the planet Mars, is the tallest known mountain in the Solar System.

There are several extinct volcanoes on Mars, four of which are vast shield volcanoes far bigger than any on Earth. They include Arsia Mons, Ascraeus Mons, Hecates Tholus,

Olympus Mons, and Pavonis Mons. These volcanoes have been extinct for many millions of years, but the European *Mars Express* spacecraft has found evidence that volcanic activity may have occurred on Mars in the recent past as well.

Jupiter's moon Io is the most volcanically active object in the solar system because of tidal interaction with Jupiter. It is covered with volcanoes that erupt sulfur, sulfur dioxide and silicate rock, and as a result, Io is constantly being resurfaced. Its lavas are the hottest known anywhere in the solar system, with temperatures exceeding 1,800 K (1,500 °C). In February 2001, the largest recorded volcanic eruptions in the solar system occurred on Io. Europa, the smallest of Jupiter's Galilean moons, also appears to have an active volcanic system, except that its volcanic activity is entirely in the form of water, which freezes into ice on the frigid surface. This process is known as cryovolcanism, and is apparently most common on the moons of the outer planets of the solar system.

In 1989 the Voyager 2 spacecraft observed cryovolcanoes (ice volcanoes) on Triton, a moon of Neptune, and in 2005 the Cassini–Huygens probe photographed fountains of frozen particles erupting from Enceladus, a moon of Saturn. The ejecta may be composed of water, liquid nitrogen, dust, or methane compounds. Cassini–Huygens also found evidence of a methane-spewing cryovolcano on the Saturnian moon Titan, which is believed to be a significant source of the methane found in its atmosphere. It is theorized that cryovolcanism may also be present on the Kuiper Belt Object Quaoar.

A 2010 study of the exoplanet COROT-7b, which was detected by transit in 2009, studied that tidal heating from the host star very close to the planet and neighboring planets could generate intense volcanic activity similar to Io.

Traditional Beliefs about Volcanoes

Many ancient accounts ascribe volcanic eruptions to supernatural causes, such as the actions of gods or demigods. To the ancient Greeks, volcanoes' capricious power could only be explained as acts of the gods, while 16th/17th-century German astronomer Johannes Kepler believed they were ducts for the Earth's tears. One early idea counter to this was proposed by Jesuit Athanasius Kircher (1602–1680), who witnessed eruptions of Mount Etna and Stromboli, then visited the crater of Vesuvius and published his view of an Earth with a central fire connected to numerous others caused by the burning of sulfur, bitumen and coal.

Various explanations were proposed for volcano behavior before the modern understanding of the Earth's mantle structure as a semisolid material was developed. For decades after awareness that compression and radioactive materials may be heat sources, their contributions were specifically discounted. Volcanic action was often attributed to chemical reactions and a thin layer of molten rock near the surface.

Lava

Lava is the molten rock expelled by a volcano during an eruption. The resulting rock after solidification and cooling is also called lava. The molten rock is formed in the interior of some planets, including Earth, and some of their satellites. The source of the heat that melts the rock within the earth is geothermal energy. When first erupted from a volcanic vent, lava is a liquid usually at temperatures from 700 to 1,200 °C (1,292 to 2,192 °F).

10-metre (33 ft) high fountain of pāhoehoe lava, Hawaii, United States

Lava flow during a rift eruption at Krafla, Iceland in 1984.

A *lava flow* is a moving outpouring of lava, which is created during a non-explosive effusive eruption. When it has stopped moving, lava solidifies to form igneous rock. The term *lava flow* is commonly shortened to *lava*. Although lava can be up to 100,000 times more viscous than water, lava can flow great distances before cooling and solidifying because of its thixotropic and shear thinning properties.

Explosive eruptions produce a mixture of volcanic ash and other fragments called tephra, rather than lava flows. The word "lava" comes from Italian, and is probably derived from the Latin word *labes* which means a fall or slide. The first use in connection with extruded magma (molten rock below the Earth's surface) was apparently in a short account written by Francesco Serao on the eruption of Vesuvius between May 14 and June 4, 1737. Serao described "a flow of fiery lava" as an analogy to the flow of water and mud down the flanks of the volcano following heavy rain.

Lava Composition

Pāhoehoe and lava flows side by side at the Big Island of Hawaii in September, 2007

The composition of almost all lava of the Earth's crust is dominated by silicate minerals, mostly feldspars, olivine, pyroxenes, amphiboles, micas and quartz.

Silicate Lavas

Igneous rocks, which form lava flows when erupted, can be classified into three chemical types; *felsic*, *intermediate*, and *mafic* (four if one includes the super-heated *ultramafic*). These classes are primarily chemical; however, the chemistry of lava also tends to correlate with the magma temperature, its viscosity and its mode of eruption.

Felsic Lava

Felsic or silicic lavas such as rhyolite and dacite typically form lava spines, lava domes or "coulees" (which are thick, short lava flows) and are associated with pyroclastic (fragmental) deposits. Most silicic lava flows are extremely viscous, and typically fragment as they extrude, producing blocky autobreccias. The high viscosity and strength are the result of their chemistry, which is high in silica, aluminium, potassium, sodium, and calcium, forming a polymerized liquid rich in feldspar and quartz, and thus has a higher viscosity than other magma types. Felsic magmas can erupt at temperatures as low as 650 to 750 °C (1,202 to 1,382 °F). Unusually hot (>950 °C; >1,740 °F) rhyolite lavas, however, may flow for distances of many tens of kilometres, such as in the Snake River Plain of the northwestern United States.

Intermediate Lava

Intermediate or andesitic lavas are lower in aluminium and silica, and usually somewhat richer in magnesium and iron. Intermediate lavas form andesite domes and block lavas, and may occur on steep composite volcanoes, such as in the Andes. Poorer in aluminium and silica than felsic lavas, and also commonly hotter (in the range of 750 to 950 °C (1,380 to 1,740 °F)), they tend to be less viscous. Greater temperatures tend to destroy polymerized bonds within the magma, promoting more fluid behaviour and also a greater tendency to form phenocrysts. Higher iron and magnesium tends to manifest as a darker groundmass, and also occasionally amphibole or pyroxene phenocrysts.

Mafic Lava

Mafic or basaltic lavas are typified by their high ferromagnesian content, and generally erupt at temperatures in excess of 950 °C (1,740 °F). Basaltic magma is high in iron and magnesium, and has relatively lower aluminium and silica, which taken together reduces the degree of polymerization within the melt. Owing to the higher temperatures, viscosities can be relatively low, although still thousands of times higher than water. The low degree of polymerization and high temperature favors chemical diffusion, so it is common to see large, well-formed phenocrysts within mafic lavas. Basalt lavas tend to produce low-profile shield volcanoes or "flood basalt fields", because the fluidal lava flows for long distances from the vent. The thickness of a basalt lava, particularly on a low slope, may be much greater than the thickness of the moving lava flow at any one time, because basalt lavas may "inflate" by supply of lava beneath a solidified crust. Most basalt lavas are of *□A□ā* or *pāhoehoe* types, rather than block lavas. Underwater they can form pillow lavas, which are rather similar to entrail-type pahoehoe lavas on land.

Ultramafic Lava

Ultramafic lavas such as komatiite and highly magnesian magmas that form boninite take the composition and temperatures of eruptions to the extreme. Komatiites contain over 18% magnesium oxide, and are thought to have erupted at temperatures of 1,600 °C (2,910 °F). At this temperature there is no polymerization of the mineral compounds, creating a highly mobile liquid with viscosity as low as that of water. Most if not all ultramafic lavas are no younger than the Proterozoic, with a few ultramafic magmas known from the Phanerozoic. No modern komatiite lavas are known, as the Earth's mantle has cooled too much to produce highly magnesian magmas.

Unusual Lavas

Some lavas of unusual composition have erupted onto the surface of the Earth. These include:

- Carbonatite and natrocarbonatite lavas are known from Ol Doinyo Lengai volcano in Tanzania, which is the sole example of an active carbonatite volcano.
- Iron oxide lavas are thought to be the source of the iron ore at Kiruna, Sweden which formed during the Proterozoic. Iron oxide lavas of Pliocene age occur at the El Laco volcanic complex on the Chile-Argentina border. Iron oxide lavas are thought to be the result of immisible separation of iron oxide magma from a parental magma of calc-alkaline or alkaline composition.

- Sulfur lava flows up to 250 metres (820 feet) long and 10 metres (33 feet) wide occur at Lastarria volcano, Chile. They were formed by the melting of sulfur deposits at temperatures as low as 113 °C (235 °F).
- Olivine nephelinite lavas are thought to have come from much deeper in the mantle of the Earth than other lavas.

The term "lava" can also be used to refer to molten "ice mixtures" in eruptions on the icy satellites of the Solar System's gas giants.

Lava Behavior

In general, the composition of a lava determines its behavior more than the temperature of its eruption. The viscosity of lava is important because it determines how the lava will behave. Lavas with high viscosity are rhyolite, dacite, andesite and trachyte, with cooled basaltic lava also quite viscous; those with low viscosities are freshly erupted basalt, carbonatite and occasionally andesite.

Toes of a pāhoehoe advance across a road in Kalapana on the east rift zone of Kīlauea Volcano in Hawaii, United States.

Highly viscous lava shows the following behaviors:

- tends to flow slowly, clog, and form semi-solid blocks which resist flow
- tends to entrap gas, which form vesicles (bubbles) within the rock as they rise to the surface
- correlates with explosive or phreatic eruptions and is associated with tuff and pyroclastic flows

Highly viscous lavas do not usually flow as liquid, and usually form explosive fragmental ash or tephra deposits. However, a degassed viscous lava or one which erupts somewhat hotter than usual may form a lava flow.

Lava with low viscosity shows the following behaviors:

- tends to flow easily, forming puddles, channels, and rivers of molten rock
- tends to easily release bubbling gases as they are formed
- eruptions are rarely pyroclastic and are usually quiescent
- volcanoes tend to form broad shields rather than steep cones

Lavas also may contain many other components, sometimes including solid crystals of various minerals, fragments of exotic rocks known as xenoliths and fragments of previously solidified lava.

Lava Morphology

The physical behavior of lava creates the physical forms of a lava flow or volcano. More fluid basaltic lava flows tend to form flat sheet-like bodies, whereas viscous rhyolite lava flows forms knobbly, blocky masses of rock.

General features of volcanology can be used to classify volcanic edifices and provide information on the eruptions which formed the lava flow, even if the sequence of lavas have been buried or metamorphosed.

Lava enters the Pacific at the Big Island of Hawaii

The ideal lava flow will have a brecciated top, either as pillow lava development, autobreccia and rubble typical of *lava* and viscous flows, or a vesicular or frothy carapace such as scoria or pumice. The top of the lava will tend to be glassy, having been flash frozen in contact with the air or water.

The centre of a lava flow is commonly massive and crystalline, flow banded or layered, with microscopic groundmass crystals. The more viscous lava forms tend to show sheeted flow features, and blocks or breccia entrained within the sticky lava. The crystal size at the centre of a lava will in general be greater than at the margins, as the crystals have more time to grow.

The base of a lava flow may show evidence of hydrothermal activity if the lava flowed across moist or wet substrates. The lower part of the lava may have vesicles, perhaps filled with minerals (amygdules). The substrate upon which the lava has flowed may show signs of scouring, it may be broken or disturbed by the boiling of trapped water, and in the case of soil profiles, may be baked into a brick-red terracotta.

Discriminating between an intrusive sill and a lava flow in ancient rock sequences can be difficult. However, some sills do not usually have brecciated margins, and may show a weak metamorphic aureole on both the upper and lower surface, whereas a lava will only bake the substrate beneath it. However, it is often difficult in practice to identify these metamorphic phenomenon because they are usually weak and restricted in size. Peperitic sills, intruded into wet sedimentary rocks, commonly do not bake upper margins and have upper and lower autobreccias, closely similar to lavas.

Lavā

Lava is one of three basic types of flow lava. Lava is basaltic lava characterized by a rough or rubbly surface composed of broken lava blocks called clinker. The Hawaiian word was introduced as a technical term in geology by Clarence Dutton.

Glowing Lava flow front advancing over pāhoehoe on the coastal plain of Kilauea in Hawaii, United States.

The loose, broken, and sharp, spiny surface of an lavā flow makes hiking difficult and slow. The clinkery surface actually covers a massive dense core, which is the most active part of the flow. As pasty lava in the core travels downslope, the clinkers are carried along at the surface. At the leading edge of an lavā flow, however, these cooled fragments tumble down the steep front and are buried by the advancing flow. This produces a layer of lava fragments both at the bottom and top of an lavā flow.

Accretionary lava balls as large as 3 metres (10 feet) are common on lavā flows. Lavā is usually of higher viscosity than pāhoehoe. Pāhoehoe can turn into lavā if it becomes turbulent from meeting impediments or steep slopes.

The sharp, angled texture makes lavā a strong radar reflector, and can easily be seen from an orbiting satellite (bright on Magellan pictures).

lavā lavas typically erupt at temperatures of 1,000 to 1,100 °C (1,830 to 2,010 °F). It originates from Hawaiian where it is pronounced, meaning "stony rough lava", but also to "burn" or "blaze".

Pāhoehoe

Pāhoehoe lava from Kīlauea volcano, Hawaii, United States

Pāhoehoe is basaltic lava that has a smooth, billowy, undulating, or ropy surface. These surface features are due to the movement of very fluid lava under a congealing surface crust. The Hawaiian word was introduced as a technical term in geology by Clarence Dutton.

A pāhoehoe flow typically advances as a series of small lobes and toes that continually break out from a cooled crust. It also forms lava tubes where the minimal heat loss maintains low viscosity. The surface texture of pāhoehoe flows varies widely, displaying all kinds of bizarre shapes often referred to as lava sculpture. With increasing distance from the source, pāhoehoe flows may change into lavā flows in response to heat loss and consequent increase in viscosity. Pahoehoe lavas typically have a temperature of 1,100 to 1,200 °C (2,010 to 2,190 °F).

Most lava flows on the Earth are less than 10 km (6.2 mi) long, but some pāhoehoe flows are more than 50 km (31 mi) long.

The rounded texture makes pāhoehoe a poor radar reflector, and is difficult to see from an orbiting satellite (dark on Magellan picture).

Block Lava Flows

Block lava flows are typical of andesitic lavas from stratovolcanoes. They behave in a similar manner to 'a' ā flows but their more viscous nature causes the surface to be covered in smooth-sided angular fragments (blocks) of solidified lava instead of clinkers. Like in 'a' ā flows, the molten interior of the flow, which is kept insulated by the solidified blocky surface, overrides the rubble that falls off the flow front. They also move much more slowly downhill and are thicker in depth than 'a' ā flows.

Domes and Coulées

Lava domes and coulées are associated with felsic lava flows ranging from dacite to rhyolite. The very viscous nature of these lava cause them to not flow far from the vent, causing the lava to form a lava dome at the vent. When a dome forms on an inclined surface it can flow in short thick flows called coulées (dome flows). These flows often travel only a few kilometers from the vent.

Pillow Lava

Pillow lava (NOAA)

Pillow lava is the lava structure typically formed when lava emerges from an underwater volcanic vent or subglacial volcano or a lava flow enters the ocean. However, pillow lava can also form when lava is erupted beneath thick glacial ice. The viscous lava gains a solid crust on contact with the water, and this crust cracks and oozes additional large blobs or "pillows" as more lava emerges from the advancing flow. Since water covers the majority of Earth's surface and most volcanoes are situated near or under bodies of water, pillow lava is very common.

Lava Landforms

Because it is formed from viscous molten rock, lava flows and eruptions create distinctive formations, landforms and topographical features from the macroscopic to the microscopic.

Volcanoes

Volcanoes are the primary landforms built by repeated eruptions of lava and ash over time. They range in shape from shield volcanoes with broad, shallow slopes formed from predominantly effusive eruptions of relatively fluid basaltic lava flows, to steeply-sided stratovolcanoes (also known as composite volcanoes) made of alternating layers of ash and more viscous lava flows typical of intermediate and felsic lavas.

Arenal Volcano, Costa Rica, is a stratovolcano.

A caldera, which is a large subsidence crater, can form in a stratovolcano, if the magma chamber is partially or wholly emptied by large explosive eruptions; the summit cone no longer supports itself and thus collapses in on itself afterwards. Such features may include volcanic crater lakes and lava domes after the event. However, calderas can also form by non-explosive means such as gradual magma subsidence. This is typical of many shield volcanoes.

Cinder and Spatter Cones

Cinder cones and spatter cones are small-scale features formed by lava accumulation around a small vent on a volcanic edifice. Cinder cones are formed from tephra or ash and tuff which is thrown from an explosive vent. Spatter cones are formed by accumulation of molten volcanic slag and cinders ejected in a more liquid form.

Kīpukas

Another Hawaiian English term derived from the Hawaiian language, a kīpuka denotes an elevated area such as a hill, ridge or old lava dome inside or downslope from an area of active volcanism. New lava flows will cover the surrounding land, isolating the kīpuka so that it appears as a (usually) forested island in a barren lava flow.

Lava Domes

Lava domes are formed by the extrusion of viscous felsic magma. They can form prominent rounded protuberances, such as at Valles Caldera. As a volcano extrudes silicic lava, it can form an *inflation dome*, gradually building up a large, pillow-like structure which cracks, fissures, and may release cooled chunks of rock and rubble. The top and side margins of an inflating lava dome tend to be covered in fragments of rock, breccia and ash.

A forested lava dome in the midst of the Valle Grande, the largest meadow in the Valles Caldera National Preserve, New Mexico, United States.

Examples of lava dome eruptions include the Novarupta dome, and successive lava domes of Mount St Helens.

Lava Tubes

Lava tubes are formed when a flow of relatively fluid lava cools on the upper surface sufficiently to form a crust. Beneath this crust, which being made of rock is an excellent insulator, the lava can continue to flow as a liquid. When this flow occurs over a prolonged period of time the lava conduit can form a tunnel-like aperture or *lava tube*, which can conduct molten rock many kilometres from the vent without cooling appreciably. Often these lava tubes drain out once the supply of fresh lava has stopped, leaving a considerable length of open tunnel within the lava flow.

Lava tubes are known from the modern day eruptions of Kīlauea, and significant, extensive and open lava tubes of Tertiary age are known from North Queensland, Australia, some extending for 15 kilometres (9 miles).

Lava Fountains

450m-high lava fountain at Kilauea

A lava fountain is a volcanic phenomenon in which lava is forcefully but non-explosively ejected from a crater, vent, or fissure. The highest lava fountains recorded were during the 1999 eruption of Mount Etna in Italy, which reached heights of 2,000 m (6,562 ft). However, lava fountains observed during Mount Vesuvius' 1779 eruption are believed to have reached at least 3,000 m (9,843 ft). Lava fountains may occur as a series of short pulses, or a continuous jet of lava. They are commonly associated with Hawaiian eruptions.

Lava Lakes

Shiprock, New Mexico, United States: a volcanic neck in the distance, with a radiating dike on its south side. Photo credit: USGS Digital Data Series

Rarely, a volcanic cone may fill with lava but not erupt. Lava which pools within the caldera is known as a lava lake. Lava lakes do not usually persist for long, either draining back into the magma chamber once pressure is relieved (usually by venting of gases through the caldera), or by draining via eruption of lava flows or pyroclastic explosion.

There are only a few sites in the world where permanent lakes of lava exist. These include:

- Mount Erebus, Antarctica
- Pu'u'Ō'ō and Halemaumau Crater on Kilauea volcano, Hawaii
- Erta Ale, Ethiopia
- Nyiragongo, Democratic Republic of Congo
- Ambrym, Vanuatu.

Lava Delta

Lava deltas form wherever sub-aerial flows of lava enter standing bodies of water. The lava cools and breaks up as it encounters the water, with the resulting fragments filling in the seabed topography such that the sub-aerial flow can move further offshore. Lava deltas are generally associated with large-scale, effusive type basaltic volcanism.

Hazards

Lava flows are enormously destructive to property in their path. However, casualties are rare since flows are usually slow enough for people and animals to escape, though this is dependent on the viscosity of the lava. Nevertheless, injuries and deaths have occurred, either because they had their escape route cut off, because they got too close to the flow or, more rarely, if the lava flow front travels too quickly. This notably happened during the eruption of Nyiragongo in Zaire (now Democratic Republic of the Congo). On the night of 10 January 1977 a crater wall was breached and a fluid lava lake drained out in under an hour. The resulting flow sped down the steep slopes at up to 100 km/h (62 mph), and overwhelmed several villages while residents were asleep. As a result of this disaster, the mountain was designated a Decade Volcano in 1991.

Deaths attributed to volcanoes frequently have a different cause, for example volcanic ejecta, pyroclastic flow from a collapsing lava dome, lahars, poisonous gases that travel ahead of lava, or explosions caused when the flow comes into contact with water. A particularly dangerous area is called a lava bench. This very young ground will typically break-off and fall into the sea.

Areas of recent lava flows continue to represent a hazard long after the lava has cooled. Where young flows have created new lands, land is more unstable and can break-off into the sea. Flows often have deep cracks, and any fall against fresh lava is similar to falling against broken glass. Rugged hiking boots, long pants, and gloves are recommended when crossing lava flows.

Towns Destroyed by Lava Flows

Lava can easily destroy entire towns. This picture shows one of over 100 houses destroyed by the lava flow in Kalapana, Hawaii, United States, in 1990.

- Kalapana, Hawaii was destroyed by the eruption of the Kīlauea volcano in 1990. (abandoned)
- Koae and Kapoho, Hawaii were both destroyed by the same eruption of Kīlauea in January, 1960. (abandoned)
- Keawaiki, Hawaii 1859 (abandoned)

- San Sebastiano al Vesuvio, Italy Destroyed in 1944 by the most recent eruption of Mount Vesuvius during the Allies' occupation of southern Italy. (rebuilt)
- Cagsawa, Philippines buried by lava erupted from Mayon Volcano in 1814.
- The Nisga'a villages of Lax Ksiluux and Wii Lax K'abit in northwestern British Columbia, Canada were destroyed by thick lava flows during the eruption of Tseax Cone in the 1700s.

Towns Damaged by Lava Flows

- Catania, Italy, in the eruption Mount Etna in 1669 (rebuilt)
- Goma, Democratic Republic of Congo, in the eruption of Nyiragongo in 2002
- Heimaey, Iceland, in the 1973 Eldfell eruption (rebuilt)
- Royal Gardens, Hawaii, by the eruption of Kilauea in 1986–87 (abandoned)
- Parícutin (village after which the volcano was named) and San Juan Parangaricutiro, Mexico, by Parícutin from 1943 to 1952.
- Sale'aula, Samoa, by eruptions of Mt Matavanu between 1905 and 1911.

Towns Destroyed by Tephra

Tephra is volcanic ash, lapilli, volcanic bombs or volcanic blocks.

- Pompeii, Italy in the eruption of Mount Vesuvius in 79 AD
- Herculaneum, Italy in the eruption of Mount Vesuvius in 79 AD
- Sumbawa Island, Indonesia in the eruption of Mount Tambora in 1815 AD
- Cerén, El Salvador in the eruption of Ilopango between 410 and 535 AD
- Plymouth, Montserrat, in 1995. Plymouth was the capital and only port of entry for Montserrat and had to be completely abandoned, along with over half of the island. It is still the *de jure* capital.

Magma

Magma is a mixture of molten or semi-molten rock, volatiles and solids that is found beneath the surface of the Earth, and is expected to exist on other terrestrial planets and some natural satel-lites. Besides molten rock, magma may also contain suspended crystals, dissolved gas and sometimes gas bubbles. Magma often collects in magma chambers that may feed a volcano or solidify underground to form an intrusion. Magma is capable of intruding into adjacent rocks (forming igneous dikes and sills), extrusion onto the surface as lava, and explosive ejection as tephra to form pyroclastic rock.

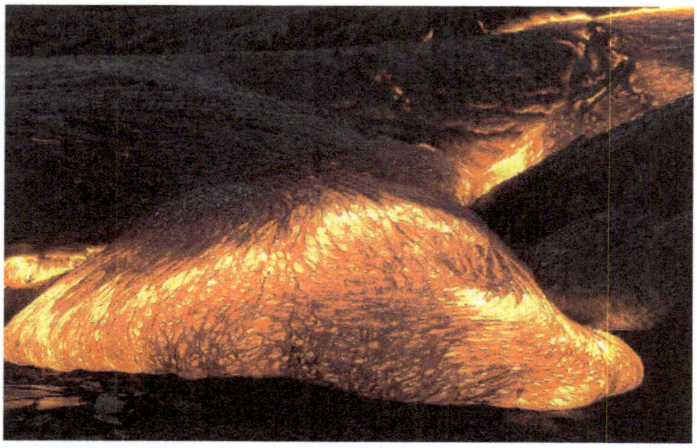

Lava flow on Hawaii. Lava is the extrusive equivalent of magma.

Description

Magma is a complex high-temperature fluid substance. Temperatures of most magmas are in the range 700 °C to 1300 °C (or 1300 °F to 2400 °F), but very rare carbonatite magmas may be as cool as 600 °C, and komatiite magmas may have been as hot as 1600 °C. Most magmas are silicate mixtures.

Environments of magma formation and compositions are commonly correlated. Environments include subduction zones, continental rift zones, mid-ocean ridges and hotspots. Despite being found in such widespread locales, the bulk of the Earth's crust and mantle is not molten. Except for the liquid outer core, most of the Earth takes the form of a rheid, a form of solid that can move or deform under pressure. Magma, as liquid, preferentially forms in high temperature, low pressure environments within several kilometers of the Earth's surface.

Magma compositions may evolve after formation by fractional crystallization, contamination, and magma mixing. By definition rock formed of solidified magma is called igneous rock.

While the study of magma has historically relied on observing magma in the form of lava outflows, magma has been encountered in situ three times during geothermal drilling projects—twice in Iceland, and once in Hawaii.

Partial Melting

Melting of solid rocks to form magma is controlled by three physical parameters: temperature, pressure, and composition. Mechanisms are discussed in the entry for igneous rock.

When rocks melt they do so incrementally and gradually; most rocks are made of several minerals, all of which have different melting points, and the physical/chemical relationships controlling melting are complex. As a rock melts, its volume changes. When enough rock is melted, the small globules of melt (generally occurring in between mineral grains) link up and soften the rock. Under pressure within the earth, as little as a fraction of a percent partial melting may be sufficient to cause melt to be squeezed from its source. Melts can stay in place long enough to melt to 20% or even 35%, but rocks are rarely melted in excess of 50%, because eventually the melted rock mass becomes a crystal and melt mush that can then ascend *en masse* as a diapir, which may then cause further decompression melting.

Geochemical Implications of Partial Melting

The degree of partial melting is critical for determining what type of magma is produced. The degree of partial melting required to form a melt can be estimated by considering the relative enrichment of incompatible elements versus compatible elements. Incompatible elements commonly include potassium, barium, cesium, and rubidium.

Rock types produced by small degrees of partial melting in the Earth's mantle are typically alkaline (Ca, Na), potassic (K) and/or peralkaline (high aluminium to silica ratio). Typically, primitive melts of this composition form lamprophyre, lamproite, kimberlite and sometimes nepheline-bearing mafic rocks such as alkali basalts and essexite gabbros or even carbonatite.

Pegmatite may be produced by low degrees of partial melting of the crust. Some granite-composition magmas are eutectic (or cotectic) melts, and they may be produced by low to high degrees of partial melting of the crust, as well as by fractional crystallization. At high degrees of partial melting of the crust, granitoids such as tonalite, granodiorite and monzonite can be produced, but other mechanisms are typically important in producing them.

Magma Usage for Energy Production

The Iceland Deep Drilling Project, while drilling several 5,000m holes in an attempt to harness the heat in the volcanic bedrock below the surface of Iceland, struck a pocket of magma at 2,100m in 2009. Being only the third time in recorded history that magma had been reached, IDDP decided to invest in the hole, naming it IDDP-1.

A cemented steel case was constructed in the hole with a perforation at the bottom close to the magma. The high temperatures and pressure of the magma steam were used to generate 36MW of power, making IDDP-1 the world's first magma-enhanced geothermal system.

Evolution of Magmas

Primary Melts

When a rock melts, the liquid is a *primary melt*. Primary melts have not undergone any differentiation and represent the starting composition of a magma. In nature it is rare to find primary melts. The leucosomes of migmatites are examples of primary melts. Primary melts derived from the mantle are especially important, and are known as *primitive melts* or primitive magmas. By finding the primitive magma composition of a magma series it is possible to model the composition of the mantle from which a melt was formed, which is important in understanding evolution of the mantle.

Parental Melts

Where it is impossible to find the primitive or primary magma composition, it is often useful to attempt to identify a parental melt. A parental melt is a magma composition from which the observed range of magma chemistries has been derived by the processes of igneous differentiation. It need not be a primitive melt.

For instance, a series of basalt flows are assumed to be related to one another. A composition from which they could reasonably be produced by fractional crystallization is termed a *parental melt*. Fractional crystallization models would be produced to test the hypothesis that they share a common parental melt.

At high degrees of partial melting of the mantle, komatiite and picrite are produced.

Migration

Magma develops within the mantle or crust when the temperature-pressure conditions favor the molten state. Magma rises toward the Earth's surface when it is less dense than the surrounding rock and when a structural zone allows movement. Magma develops or collects in areas called magma chambers. Magma can remain in a chamber until it cools and crystallizes forming igneous rock, it erupts as a volcano, or moves into another magma chamber.

Cooling of Magmas

There are two known processes by which magma ceases to exist: by volcanic eruption, or by crystallization within the crust or mantle to form a pluton. In both cases the bulk of the magma eventually cools and forms igneous rocks.

When magma cools it begins to form solid mineral phases. Some of these settle at the bottom of the magma chamber forming cumulates that might form mafic layered intrusions. Magma that cools slowly within a magma chamber usually ends up forming bodies of plutonic rocks such as gabbro, diorite and granite, depending upon the composition of the magma. Alternatively, if the magma is erupted it forms volcanic rocks such as basalt, andesite and rhyolite (the extrusive equivalents of gabbro, diorite and granite, respectively).

Volcanism

During a volcanic eruption the magma that leaves the underground is called lava. Lava cools and solidifies relatively quickly compared to underground bodies of magma. This fast cooling does not allow crystals to grow large, and a part of the melt does not crystallize at all, becoming glass. Rocks largely composed of volcanic glass include obsidian, scoria and pumice.

Before and during volcanic eruptions, volatiles such as CO_2 and H_2O partially leave the melt through a process known as exsolution. Magma with low water content becomes increasingly viscous. If massive exsolution occurs when magma heads upwards during a volcanic eruption, the resulting eruption is usually explosive.

Composition, Melt Structure and Properties

Silicate melts are composed mainly of silicon, oxygen, aluminium, alkalis (sodium, potassium, calcium), magnesium and iron. Silicon atoms are in tetrahedral coordination with oxygen, as in almost all silicate minerals, but in melts atomic order is preserved only over short distances. The physical behaviours of melts depend upon their atomic structures as well as upon temperature and pressure and composition.

Viscosity is a key melt property in understanding the behaviour of magmas. More silica-rich melts are typically more polymerized, with more linkage of silica tetrahedra, and so are more viscous. Dissolution of water drastically reduces melt viscosity. Higher-temperature melts are less viscous.

Generally speaking, more mafic magmas, such as those that form basalt, are hotter and less viscous than more silica-rich magmas, such as those that form rhyolite. Low viscosity leads to gentler, less explosive eruptions.

Characteristics of several different magma types are as follows:

Ultramafic (picritic)

$SiO_2 < 45\%$

Fe–Mg > 8% up to 32%MgO

Temperature: up to 1500°C

Viscosity: Very Low

Eruptive behavior: gentle or very explosive (kimberilites)

Distribution: divergent plate boundaries, hot spots, convergent plate boundaries; komatiite and other ultramafic lavas are mostly Archean and were formed from a higher geothermal gradient and are unknown in the present

Mafic (basaltic)

SiO_2 < 50%

FeO and MgO typically < 10 wt%

Temperature: up to ~1300°C

Viscosity: Low

Eruptive behavior: gentle

Distribution: divergent plate boundaries, hot spots, convergent plate boundaries

Intermediate (andesitic)

SiO_2 ~ 60%

Fe–Mg: ~ 3%th

Temperature: ~1000°C

Viscosity: Intermediate

Eruptive behavior: explosive or effusive

Distribution: convergent plate boundaries, island arcs

Felsic (rhyolitic)

SiO_2 > 70%

Fe–Mg: ~ 2%

Temp: < 900°C

Viscosity: High

Eruptive behavior: explosive or effusive

Distribution: common in hot spots in continental crust (Yellowstone National Park) and in continental rifts

Temperature

At any given pressure and for any given composition of rock, a rise in temperature past the solidus will cause melting. Within the solid earth, the temperature of a rock is controlled by the geothermal gradient and the radioactive decay within the rock. The geothermal gradient averages about 25 °C/km with a wide range from a low of 5–10 °C/km within oceanic trenches and subduction zones to 30–80 °C/km under mid-ocean ridges and volcanic arc environments.

Pressure

As magma buoyantly rises it will cross the solidus-liquidus and its temperature will reduce by adiabatic cooling. At this point it will liquefy and if erupted onto the surface will form lava. Melting can also occur due to a reduction in pressure by a process known as decompression melting.

Density

Type	Density [kg/m³]
Basalt magma	2650–2800
Andesite magma	2450–2500
Rhyolite magma	2180–2250

Composition

It is usually very difficult to change the bulk composition of a large mass of rock, so composition is the basic control on whether a rock will melt at any given temperature and pressure. The composition of a rock may also be considered to include *volatile* phases such as water and carbon dioxide.

The presence of volatile phases in a rock under pressure can stabilize a melt fraction. The presence of even 0.8% water may reduce the temperature of melting by as much as 100 °C. Conversely, the loss of water and volatiles from a magma may cause it to essentially freeze or solidify.

Also a major portion of almost all magma is silica, which is a compound of silicon and oxygen. Magma also contains gases, which expand as the magma rises. Magma that is high in silica resists flowing, so expanding gases are trapped in it. Pressure builds up until the gases blast out in a violent, dangerous explosion. Magma that is relatively poor in silica flows easily, so gas bubbles move up through it and escape fairly gently.

Important Aspect of Magma

Magma Chamber

A magma chamber is a large underground pool of liquid rock found beneath the surface of the Earth. The molten rock in such a chamber is under great pressure, and given enough time, that pressure can gradually fracture the rock around it creating outlets for the magma. If it finds a way to the surface, then the result will be a volcanic eruption; consequently many volcanoes are situated over magma chambers. Magma chambers are hard to detect, and most of the known ones are therefore close to the surface of the Earth, commonly between 1 km and 10 km under the surface. In geological terms, this is extremely close to the surface, although in human terms, it is considerably deep underground.

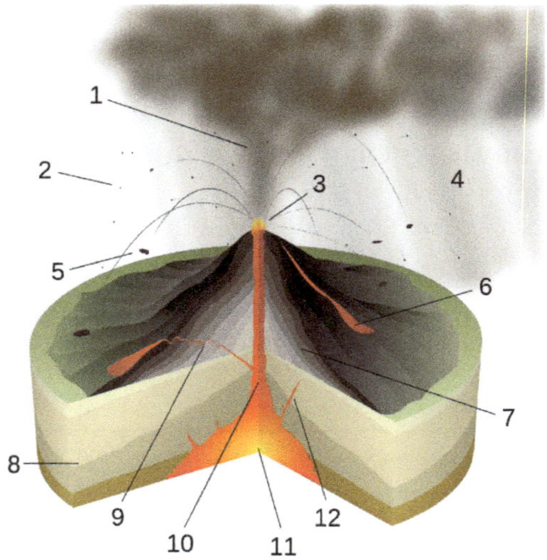

11 – magma chamber

Dynamics of Magma Chambers

Magma rises through cracks from beneath and across the crust because it is less dense than the surrounding rock. When the magma cannot find a path upwards it pools into a magma chamber. Magma chambers are commonly built up incrementally, by successive horizontal or vertical magma injections. Influx of new magma causes reaction of pre-existing crystals and the pressure in the chamber to increase.

The residing magma starts to cool, with the higher melting point components such as olivine crystallizing out of the solution, particularly near to the cooler walls of the chamber, and forming a denser conglomerate of minerals which sinks (cumulative rock). Upon cooling, new mineral phases saturate and the rock type changes (e.g. fractional crystallization), typically forming gabbro, diorite, tonalite and granite or gabbro, diorite, syenite and granite. If magma resides in a chamber for a long period, then it can become stratified with lower density components rising to the top and denser materials sinking. Rocks accumulate in layers, forming a layered intrusion. Any subsequent eruption may produce distinctly layered deposits, for example the deposits from the 79 AD eruption of Mount Vesuvius include a thick layer of white pumice from the upper portion of the magma chamber overlaid with a similar layer of grey pumice produced from material erupted later from lower down in the chamber.

Another effect of the cooling of the chamber is that the solidifying crystals will release the gas (primarily steam) previously dissolved when they were liquid, causing the pressure in the chamber to rise, possibly sufficiently to produce an eruption. Additionally, the removal of the lower melting point components will tend to make the magma more viscous (by increasing the concentration of silicates). Thus, stratification of a magma chamber may result in an increase in the amount of gas within the magma near the top of the chamber, and also make this magma more viscous; potentially leading to a more explosive eruption than would be the case had the chamber not become stratified.

If the magma is not vented to the surface in a volcanic eruption, it will slowly cool and crystallize at depth to form an intrusive igneous body composed of granite or gabbro

Often, a volcano may have a deep magma chamber many kilometers down, which supplies a shallower chamber near the summit. The location of magma chambers can be mapped using seismology: seismic waves from earthquakes move more slowly through liquid rock than solid, allowing measurements to pinpoint the regions of slow movement which identify magma chambers.

As a volcano erupts, emptying the magma chamber, the surrounding rock will collapse into it. If a large amount of magma is erupted, causing the chamber to reduce considerably in volume, then this can result in a depression at the surface called a caldera.

Tephra

Tephra is fragmental material produced by a volcanic eruption regardless of composition, fragment size or emplacement mechanism.

Volcanic tephra at Brown Bluff, Antarctica (2016)

Volcanologists also refer to airborne fragments as pyroclasts. Once clasts have fallen to the ground they remain as tephra unless hot enough to fuse together into pyroclastic rock or tuff.

Overview

The distribution of tephra following an eruption usually involves the largest boulders falling to the ground quickest and therefore closest to the vent, while smaller fragments travel further — ash can often travel for thousands of miles, even circumglobal, as it can stay in the stratosphere for days to weeks following an eruption. When large amounts of tephra accumulate in the atmosphere from massive volcanic eruptions (or from a multitude of smaller eruptions occurring simultaneously), they can reflect light and heat from the sun back through the atmosphere, in some cases causing the temperature to drop, resulting in a temporary, "volcanic winter", climate change. Tephra mixed in with precipitation can also be acidic and cause acid rain and snowfall.

Rocks from the Bishop tuff, uncompressed with pumice on left; compressed with fiamme on right.

Classification

Tephra fragments are classified by size:

Volcanic breccia in Jackson Hole.

- Ash – particles smaller than 2 mm (0.08 inches) in diameter,
- Lapilli or volcanic cinders – between 2 and 64 mm (0.08 and 2.5 inches) in diameter,
- Volcanic bombs or volcanic blocks – larger than 64 mm (2.5 inches) in diameter.

The use of tephra layers, which bear their own unique chemistry and character, as temporal marker horizons in archaeological and geological sites is known as tephrochronology.

Etymology

The word "tephra" and "pyroclast" both derive from *tephra* means "ash", while the word *pyroclast* is derived from (*pyr*), meaning "fire", and (*klastos*), meaning "broken in pieces".

Volcanic Ash

Volcanic ash consists of fragments of pulverized rock, minerals and volcanic glass, created during volcanic eruptions and measuring less than 2 mm (0.079 inches) in diameter. The term volcanic ash is also often loosely used to refer to all explosive eruption products (correctly referred to as tephra), including particles larger than 2mm. Volcanic ash is formed during explosive volcanic eruptions when dissolved gases in magma expand and escape violently into the atmosphere. The force of the escaping gas shatters the magma and propels it into the atmosphere where it solidifies into fragments of volcanic rock and glass. Ash is also produced when magma comes into contact with water during phreatomagmatic eruptions, causing the water to explosively flash to steam leading to shattering of magma. Once in the air, ash is transported by wind up to thousands of kilometers away.

Ash cloud from the 2008 eruption of Chaitén volcano, Chile, stretching across Patagonia from the Pacific to the Atlantic Ocean.

Due to its wide dispersal, ash can have a number of impacts on society, including human and animal health, disruption to aviation, disruption to critical infrastructure (e.g., electric power supply systems, telecommunications, water and waste-water networks, transportation), primary industries (e.g., agriculture), buildings and structures.

Formation

Volcanic ash is formed during explosive volcanic eruptions, phreatomagmatic eruptions and during transport in pyroclastic density currents.

Explosive eruptions occur when magma decompresses as it rises, allowing dissolved volatiles (dominantly water and carbon dioxide) to exsolve into gas bubbles. As more bubbles nucleate a foam is produced, which decreases the density of the magma, accelerating it up the conduit. Fragmentation occurs when bubbles occupy ~70-80 vol% of the erupting mixture. When fragmentation occurs, violently expanding bubbles tear the magma apart into fragments which are ejected into the atmosphere where they solidify into ash particles. Fragmentation is a very efficient process of ash formation and is capable of generating very fine ash even without the addition of water.

454 million year old volcanic ash between layers of limestone in the catacombs of Peter the Great's Naval Fortress in Estonia near Laagri. The diameter of objective cover is 58 mm (2.3 in). This is remnant of one of the oldest large eruptions preserved.

Volcanic ash is also produced during phreatomagmatic eruptions. During these eruptions fragmentation occurs when magma comes into contact with bodies of water (such as the sea, lakes and marshes) groundwater, snow or ice. As the magma, which is significantly hotter than the boiling point of water, comes into contact with water an insulating vapor film forms (Leidenfrost effect). Eventually this vapor film will collapse leading to direct coupling of the cold water and hot magma. This increases the heat transfer which leads to the rapid expansion of water and fragmentation of the magma into small particles which are subsequently ejected from the volcanic vent. Fragmentation causes an increase in contact area between magma and water creating a feedback mechanism, leading to further fragmentation and production of fine ash particles.

Pyroclastic density currents can also produce ash particles. These are typically produced by lava dome collapse or collapse of the eruption column. Within pyroclastic density currents particle abrasion occurs as particles interact with each other resulting in a reduction in grain size and production of fine grained ash particles. In addition, ash can be produced during secondary fragmentation of pumice fragments, due to the conservation of heat within the flow. These processes produce large quantities of very fine grained ash which is removed from pyroclastic density currents in co-ignimbrite ash plumes.

Physical and chemical characteristics of volcanic ash are primarily controlled by the style of volcanic eruption. Volcanoes display a range of eruption styles which are controlled by magma chemistry, crystal content, temperature and dissolved gases of the erupting magma and can be classified using the volcanic explosivity index (VEI). Effusive eruptions (VEI 1) of basaltic composition produce $<10^5$ m^3 of ejecta, whereas extremely explosive eruptions (VEI 5+) of rhyolitic and dacitic composition can inject large quantities ($>10^9$ m^3) of ejecta into the atmosphere. Another parameter controlling the amount of ash produced is the duration of the eruption: the longer the eruption is sustained, the more ash will be produced. For example, the second phase of the 2010 eruptions of Eyjafjallajökull was classified as VEI 4 despite a modest 8 km high eruption column, but the eruption continued for a month, which allowed a large volume of ash to be ejected into the atmosphere.

Properties

Chemical

The types of minerals present in volcanic ash are dependent on the chemistry of the magma from which it erupted. Considering that the most abundant elements found in magma are silica (SiO_2) and oxygen, the various types of magma (and therefore ash) produced during volcanic eruptions are most commonly explained in terms of their silica content. Low energy eruptions of basalt produce a characteristically dark coloured ash containing ~45 - 55% silica that is generally rich in iron (Fe) and magnesium (Mg). The most explosive rhyolite eruptions produce a felsic ash that is high in silica (>69%) while other types of ash with an intermediate composition (e.g., andesite or dacite) have a silica content between 55-69%.

The principal gases released during volcanic activity are water, carbon dioxide, sulfur dioxide, hydrogen, hydrogen sulfide, carbon monoxide and hydrogen chloride. These sulfur and halogen gases and metals are removed from the atmosphere by processes of chemical reaction, dry and wet deposition, and by adsorption onto the surface of volcanic ash.

It has long been recognised that a range of sulfate and halide (primarily chloride and fluoride) compounds are readily mobilised from fresh volcanic ash.; It is considered most likely that these salts are formed as a consequence of rapid acid dissolution of ash particles within eruption plumes, which is thought to supply the cations involved in the deposition of sulfate and halide salts.

While some 55 ionic species have been reported in fresh ash leachates the most abundant species usually found are the cations Na^+, K^+, Ca^{2+} and Mg^{2+} and the anions Cl^-, F^- and SO_4^{2-}. Molar ratios between ions present in leachates suggest that in many cases these elements are present as simple salts such as NaCl and $CaSO_4$. In a sequential leaching experiment on ash from the 1980 eruption of Mount St. Helens, chloride salts were found to be the most readily soluble, followed by sulfate salts Fluoride compounds are in general only sparingly soluble (e.g., CaF_2, MgF_2), with the exception of fluoride salts of alkali metals and compounds such as calcium hexafluorosilicate ($CaSiF_6$). The pH of fresh ash leachates is highly variable, depending on the presence of an acidic gas condensate (primarily as a consequence of the gases SO_2, HCl and HF in the eruption plume) on the ash surface.

The crystalline-solid structure of the salts act more as an insulator than a conductor. However, once the salts are dissolved into a solution by a source of moisture (e.g., fog, mist, light rain, etc.), the ash may become corrosive and electrically conductive. A recent study has shown that the electrical conductivity of volcanic ash increases with (1) increasing moisture content, (2) increasing soluble salt content, and (3) increasing compaction (bulk density). The ability of volcanic ash to conduct electric current has significant implications for electric power supply systems.

Physical

Components

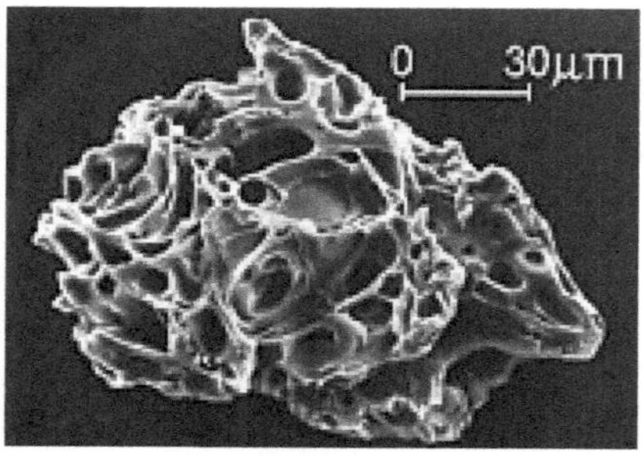

Particle of volcanic ash from Mount St. Helens.

Volcanic ash particles erupted during magmatic eruptions are made up of various fractions of vitric (glassy, non-crystalline), crystalline or lithic (non-magmatic) particles. Ash produced during low viscosity magmatic eruptions (e.g., Hawaiian and Strombolian basaltic eruptions) produce a range of different pyroclasts dependent on the eruptive process. For example, ash collected from Hawaiian lava fountains consists of sideromelane (light brown basaltic glass) pyroclasts which contain rare microlites (small quench crystals) and phenocrysts. Slightly more viscous eruptions of basalt (e.g., Strombolian) form a variety of pyroclasts from irregular sideromelane droplets to blocky tachylite (black to dark brown microcrystalline pyroclasts). In contrast, most high-silica ash (e.g. rhyolite) consists of pulverised products of pumice (vitric shards), individual phenocrysts (crystal fraction) and some lithic fragments (xenoliths).

Ash generated during phreatic eruptio[e]ns primarily consists of hydrothermally altered lithic and mineral fragments, commonly in a clay matrix. Particle surfaces are often coated with aggregates of zeolite crystals or clay and only relict textures remain to identify pyroclast types.

Morphology

The morphology (shape) of volcanic ash is controlled by a plethora of different eruption and kinematic processes. Eruptions of low-viscosity magmas (e.g., basalt) typically form droplet shaped particles. This droplet shape is, in part, controlled by surface tension, acceleration of the droplets after they leave the vent, and air friction. Shapes range from perfect spheres to a variety of twisted, elongate droplets with smooth, fluidal surfaces.

Light microscope image of ash from the 1980 eruption of Mount St. Helens, Washington.

The morphology of ash from eruptions of high-viscosity magmas (e.g., rhyolite, dacite, and some andesites) is mostly dependent on the shape of vesicles in the rising magma before disintegration. Vesicles are formed by the expansion of magmatic gas before the magma has solidified. Ash particles can have varying degrees of vesicularity and vesicular particles can have extremely high surface area to volume ratios. Concavities, troughs, and tubes observed on grain surfaces are the result of broken vesicle walls. Vitric ash particles from high-viscosity magma eruptions are typically angular, vesicular pumiceous fragments or thin vesicle-wall fragments while lithic fragments in volcanic ash are typically equant, or angular to subrounded. Lithic morphology in ash is generally controlled by the mechanical properties of the wall rock broken up by spalling or explosive expansion of gases in the magma as it reaches the surface.

The morphology of ash particles from phreatomagmatic eruptions is controlled by stresses within the chilled magma which result in fragmentation of the glass to form small blocky or pyramidal glass ash particles. Vesicle shape and density play only a minor role in the determination of grain shape in phreatomagmatic eruptions. In this sort of eruption, the rising magma is quickly cooled on contact with ground or surface water. Stresses within the "quenched" magma cause fragmentation into five dominant pyroclast shape-types: (1) blocky and equant; (2) vesicular and irregular with smooth surfaces; (3) moss-like and convoluted; (4) spherical or drop-like; and (5) plate-like.

Density

The density of individual particles varies with different eruptions. The density of volcanic ash varies between 700–1200 kg/m^3 for pumice, 2350–2450 kg/m^3 for glass shards, 2700–3300 kg/m^3 for crystals, and 2600–3200 kg/m^3 for lithic particles. Since coarser and denser particles are deposited close to source, fine glass and pumice shards are relatively enriched in ash fall deposits at distal locations. The high density and hardness (~5 on the Mohs Hardness Scale) together with a high degree of angularity, make some types of volcanic ash (particularly those with a high silica content) very abrasive.

Grain Size

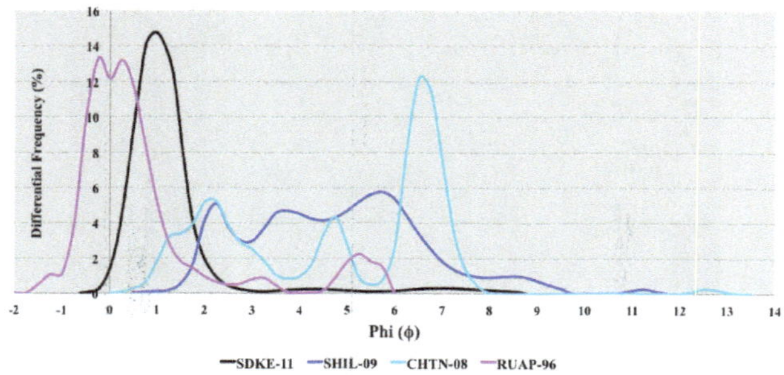

Volcanic ash grain size distributions.

Volcanic ash consists of particles (pyroclasts) with diameters <2 mm (particles >2 mm are classified as lapilli), and can be as fine as 1 μm. The overall grain size distribution of ash can vary greatly with different magma compositions. Few attempts have been made to correlate the grain size characteristics of a deposit with those of the event which produced it, though some predictions can be made. Rhyolitic magmas generally produce finer grained material compared to basaltic magmas, due to the higher viscosity and therefore explosivity. The proportions of fine ash are higher for silicic explosive eruptions, probably because vesicle size in the pre-eruptive magma is smaller than those in mafic magmas. There is good evidence that pyroclastic flows produce high proportions of fine ash by communition and it is likely that this process also occurs inside volcanic conduits and would be most efficient when the magma fragmentation surface is well below the summit crater.

Dispersal

Ash plume rising from Mount Redoubt after an eruption on April 21, 1990.

Ash particles are incorporated into eruption columns as they are ejected from the vent at high velocity. The initial momentum from the eruption propels the column upwards. As air is drawn into the column, the bulk density decreases and it starts to rise buoyantly into the atmosphere. At a point where the bulk density of the column is the same as the surrounding atmosphere, the column will cease rising and start moving laterally. Lateral dispersion is controlled by prevailing winds and the ash may be deposited hundreds to thousands of kilometres from the volcano, depending on eruption column height, particle size of the ash and climatic conditions (especially wind direction and strength and humidity).

Ash fallout occurs immediately after the eruption and is controlled by particle density. Initially, coarse particles fall out close to source. This is followed by fallout of accretionary lapilli, which is the result of particle agglomeration within the column. Ash fallout is less concentrated during the final stages as the column moves downwind. This results in an ash fall deposit which generally decreases in thickness and grain size exponentially with increasing distance from the volcano. Fine ash particles may remain in the atmosphere for days to weeks and be dispersed by high-altitude winds. These particles can impact on the aviation industry (refer to impacts section) and, combined with gas particles, can affect global climate.

Volcanic ash plumes can form above pyroclastic density currents, these are called co-ignimbrite plumes. As pyroclastic density currents travel away from the volcano, smaller particles are removed from the flow by elutriation and form a less dense zone overlying the main flow. This zone then entrains the surrounding air and a buoyant co-ignimbrite plume is formed. These plumes tend to have higher concentrations of fine ash particles compared to magmatic eruption plumes due to the abrasion within the pyroclastic density current.

Impacts

Introduction

Population growth has caused the progressive encroachment of urban development into higher risk areas, closer to volcanic centres, increasing the human exposure to volcanic ash fall events.

Infrastructure is critical to supporting modern societies, particularly in urban areas, where high population densities create high demand for services. These infrastructure networks and systems support urban living, and provide lifeline services upon which we depend for our health, education, transport and social networking. Infrastructure networks and services support a variety of facilities across a broad range of sectors.

Volcanic ash fall events can disrupt and or damage the infrastructure upon which society depends. Several recent eruptions have illustrated the vulnerability of urban areas that received only a few millimetres or centimetres of volcanic ash. This has been sufficient to cause disruption of transportation, electricity, water, sewage and storm water systems. Costs have been incurred from business disruption, replacement of damaged parts and insured losses. Ash fall impacts on critical infrastructure can also cause multiple knock-on effects, which may disrupt many different sectors and services.

Volcanic ash fall is physically, socially and economically disruptive. Volcanic ash can affect both proximal areas and areas many hundreds of kilometres from the source, and causes disruptions and losses in a wide variety of different infrastructure sectors. Impacts are dependent on: ash fall thickness; the duration of the ash fall; the grain size and chemistry of the ash; whether the ash is wet or dry; and any preparedness, management and prevention (mitigation) measures employed to reduce effects from the ash fall. Different sectors of infrastructure and society are affected in different ways and are vulnerable to a range of impacts or consequences. These are discussed in the following sections.

Infrastructure Sectors

Electricity

Electrical insulator flashover caused by volcanic ash contamination.

Volcanic ash can cause disruption to electric power supply systems at all levels of power generation, transformation, transmission and distribution. There are four main impacts arising from ash-contamination of apparatus used in the power delivery process:

- Wet deposits of ash on high voltage insulators can initiate a leakage current (small amount of current flow across the insulator surface) which, if sufficient current is achieved, can cause 'flashover' (the unintended electrical discharge around or over the surface of an insulating material).

If the resulting short-circuit current is high enough to trip the circuit breaker then disruption of service will occur. Ash-induced flashover across transformer insulation (bushings) can burn, etch or crack the insulation irreparably and will likely result in the disruption of power supply.

- Volcanic ash can erode, pit and scour metallic apparatus, particularly moving parts such as water and wind turbines and cooling fans on transformers or thermal power plants.
- The high bulk density of some ash deposits can cause line breakage and damage to steel towers and wooden poles due to ash loading. This is most hazardous when the ash and/or the lines and structures are wet (e.g., by rainfall) and there has been ≥10 mm of ash fall. Fine-grained ash (e.g., <0.5 mm diameter) adheres to lines and structures most readily. Volcanic ash may also load overhanging vegetation, causing it to fall onto lines. Snow and ice accumulation on lines and overhanging vegetation further increases the risk of breakage and or collapse of lines and other hardware.
- Controlled outages of vulnerable connection points (e.g., substations) or circuits until ash fall has subsided or for de-energised cleaning of equipment.

Drinking Water Supplies

Following an eruption, it is very common for the public to hold fears about chemical contamination of water supplies. However, in general, the physical impacts of an ashfall will tend to overwhelm problems caused by the release of chemical contaminants from fresh volcanic ash. Impacts vary according to the type of treatment system.

Large Water Treatment Plants

Groundwater-fed systems are resilient to impacts from ashfall, although airborne ash can interfere with the operation of well-head pumps. Electricity outages caused by ashfall can also disrupt electrically powered pumps if there is no backup generation.

Water turbine from the Agoyan Hydroelectric plant eroded by volcanic ash laden water.

For surface water sources such as lakes and reservoirs, the volume available for dilution of ionic species leached from ash is generally large. The most abundant components of ash leachates (Ca, Na, Mg, K, Cl, F and SO_4) occur naturally at significant concentrations in most surface waters and therefore are not affected greatly by inputs from volcanic ashfall, and are also of low concern in drinking water, with the possible exception of fluorine. The elements iron, manganese and aluminium are commonly enriched over background levels by volcanic ashfall. These elements may impart a metallic taste to water, and may produce red, brown or black staining of whiteware, but are not considered a health risk. Volcanic ashfalls are not known to have caused problems in water supplies for toxic trace elements such as mercury (Hg) and lead (Pb) which occur at very low levels in ash leachates.

A further point to note is that drinking-water treatment commonly involves the addition of treatment chemicals such as aluminium sulfate or ferric chloride as flocculants, lime for pH adjustment, chlorine for disinfection and fluoride compounds for dental health.

The physical impacts of ashfall can affect the operation of water treatment plants. Ash can block intake structures, cause severe abrasion damage to pump impellers and overload pump motors. Many water treatment plants have an initial coagulation/flocculation step that is automatically adjusted to turbidity (the level of suspended solids, measured in nephelometric turbidity units) in the incoming water. In most cases, changes in turbidity caused by suspended ash particles will be within the normal operating range of the plant and can be managed satisfactorily by adjusting the addition of coagulant. Ashfalls will be more likely to cause problems for plants that are not designed for high levels of turbidity and which may omit coagulation/flocculation treatment. Ash can enter filtration systems such as open sand filters both by direct fallout and via intake waters. In most cases, increased maintenance will be required to manage the effects of an ashfall, but there will not be service interruptions.

The final step of drinking water treatment is disinfection to ensure that final drinking water is free from infectious microorganisms. As suspended particles (turbidity) can provide a growth substrate for microorganisms and can protect them from disinfection treatment, it is extremely important that the water treatment process achieves a good level of removal of suspended particles.

Small Treatment Systems

Many small communities obtain their drinking water from diverse sources (lakes, streams, springs and groundwater wells). Levels of treatment vary widely, from rudimentary systems with coarse screening or settling followed by disinfection (usually chlorination), to more sophisticated systems using a filtration step. It should be noted that unless a high quality source is used, such as secure groundwater, disinfection alone is unlikely to guarantee that drinking water is safe from protozoa such as Giardia and

Cryptosporidium, which are relatively resistant to standard disinfectants and which require additional removal steps such as filtration.

Volcanic ashfall is likely to have major effects on these systems. Ash will clog intake structures, cause abrasion damage to pumps and block pipes, settling ponds and open filters. High levels of turbidity are very likely to interfere with disinfection treatment and doses may have to be adjusted to compensate. It is essential to monitor chlorine residuals in the distribution system.

Rainwater-fed Supplies

Many households, and some small communities, rely on rainwater for their drinking water supplies. Roof-fed systems are highly vulnerable to contamination by ashfall, as they have a large surface area relative to the storage tank volume. In these cases, leaching of chemical contaminants from the ashfall can become a health risk and drinking of water is not recommended. Prior to an ashfall, downpipes should be disconnected so that water in the tank is protected. A further problem is that the surface coating of fresh volcanic ash can be acidic. Unlike most surface waters, rainwater generally has a very low alkalinity (acid-neutralising capacity) and thus ashfall may acidify tank waters. This may lead to problems with plumbosolvency, whereby the water is more aggressive towards materials that it comes into contact with. This can be a particular problem if there are lead-head nails or lead flashing used on the roof, and for copper pipes and other metallic plumbing fittings.

Water Demand

During ashfall events large demands are commonly placed on water resources for cleanup and shortages can result. Shortages compromise key services such as firefighting and can lead to a lack of water for hygiene, sanitation and drinking. Municipal authorities need to monitor and manage this water demand carefully, and may need to advise the public to utilise cleanup methods that do not use water (e.g., cleaning with brooms rather than hoses).

Wastewater Treatment

Wastewater networks may sustain damage similar to water supply networks. It is very difficult to exclude ash from the sewerage system. Systems with combined storm water/sewer lines are most at risk. Ash will enter sewer lines where there is inflow/infiltration by stormwater through illegal connections (e.g., from roof downpipes), cross connections, around manhole covers or through holes and cracks in sewer pipes.

Ash-laden sewage entering a treatment plant is likely to cause failure of mechanical prescreening equipment such as step screens or rotating screens. Ash that penetrates further into the system will settle and reduce the capacity of biological reactors as well as increasing the volume of sludge and changing its composition.

Aircraft

The principal damage sustained by aircraft flying into a volcanic ash cloud is abrasion to forward-facing surfaces, such as the windshield and leading edges of the wings, and accumulation of ash into surface openings, including engines. Abrasion of windshields and landing lights will reduce visibility forcing pilots to rely on their instruments. However, some instruments may provide incorrect readings as sensors (e.g., pitot tubes) can become blocked with ash. Ingestion of ash into engines causes abrasion damage to compressor fan blades. The ash erodes sharp blades in the compressor, reducing its efficiency. The ash melts in the combustion chamber to form molten glass. The ash then solidifies on turbine blades, blocking air flow and causing the engine to stall.

The composition of most ash is such that its melting temperature is within the operating temperature (>1000 °C) of modern large jet engines. The degree of impact depends upon the concentration of ash in the plume, the length of time the aircraft spends within the plume and the actions taken by the pilots. Critically, melting of ash, particularly volcanic glass, can result in accumulation of resolidified ash on turbine nozzle guide vanes, resulting in compressor stall and complete loss of engine thrust. The standard procedure of the engine control system when it detects a possible stall is to increase power which would exacerbate the problem. It is recommended that pilots reduce engine power and quickly exit the cloud by performing a descending 180° turn. Volcanic gases, which are present within ash clouds, can also cause damage to engines and acrylic windshields, although this damage may not surface for many years.

Occurrence

There are many instances of damage to jet aircraft as a result of an ash encounter. On 24 June 1982 a British Airways Boeing 747-236B (Flight 9) flew through the ash cloud from the eruption of Mount Galunggung, Indonesia resulting in the failure of all four engines. The plane descended 24,000 feet (7,300 m) in 16 minutes before the engines restarted, allowing the aircraft to make an emergency landing. On 15 December 1989 a KLM Boeing 747-400 (Flight 867) also lost power to all four engines after flying into an ash cloud from Mount Redoubt, Alaska. After dropping 14,700 feet (4,500 m) in four minutes, the engines were started just 1–2 minutes before impact. Total damage was US$80 million and it took 3 months' work to repair the plane. In the 1990s a further US$100 million of damage was sustained by commercial aircraft (some in the air, others on the ground) as a consequence of the 1991 eruption of Mount Pinatubo in the Philippines.

In April 2010 airspace all over Europe was affected, with many flights cancelled-which was unprecedented-due to the presence of volcanic ash in the upper atmosphere from the eruption of the Icelandic volcano Eyjafjallajökull. On 15 April 2010 the Finnish Air Force halted training flights when damage was found from volcanic dust ingestion by the engines of one of its Boeing F-18 Hornet fighters. On 22 April 2010 UK RAF Typhoon training flights were also temporarily suspended after deposits of volcanic ash were found in a jet's engines. In June 2011 there were similar closures of airspace in Chile, Argentina, Brazil, Australia and New Zealand, following the eruption of Puyehue-Cordón Caulle, Chile.

Detection

Volcanic ash clouds are very difficult to detect from aircraft as no onboard cockpit instruments exist to detect them. However, a new system called Airborne Volcanic Object Infrared Detector (AVOID) has recently been developed by Nicarnica Aviation, a daughter company of the Norwegian Institute for Air Research, which will allow pilots to detect ash plumes up to 100 km (62 mi) ahead and fly safely around them. The system uses two fast-sampling infrared cameras, mounted on a forward-facing surface, that are tuned to detect volcanic ash. This system can detect ash concentrations of <1 mg/m³ to > 50 mg/m³, giving pilots approximately 7–10 minutes warning. The camera is currently being tested by easyJet airline company.

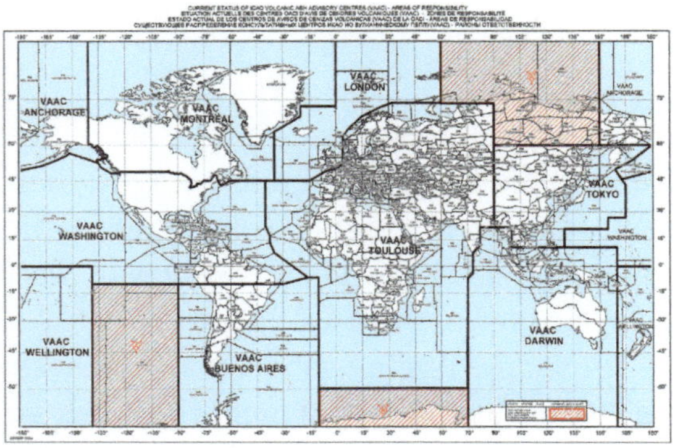

Coverage of the nine VAAC around the world

In addition, ground and satellite based imagery, radar, and lidar can be used to detect ash clouds. This information is passed between meteorological agencies, volcanic observatories and airline companies through Volcanic Ash Advisory Centers (VAAC). There is one VAAC for each of the nine regions of the world. VAACs can issue advisories describing the current and future extent of the ash cloud.

Airport Systems

Volcanic ash not only affects in-flight operations but can affect ground-based airport operations as well. Small accumulations of ash can reduce visibility, create slippery runways and taxiways, infiltrate communication and electrical systems, interrupt ground services, damage buildings and parked aircraft. Ash accumulation of more than a few millimeters requires removal before airports can resume full operations. Ash does not disappear (unlike snowfalls) and must be disposed of in a manner that prevents it from being remobilised by wind and aircraft.

Land Transport

Ash may disrupt transportation systems over large areas for hours to days, including roads and vehicles, railways and ports and shipping. Falling ash will reduce the visibility which can make driving difficult and dangerous. In addition, fast travelling cars will stir up ash, creating billowing clouds which perpetuate ongoing visibility hazards. Ash accumulations will decrease traction, especially when wet, and cover road markings. Fine-grained ash can infiltrate openings in cars and abrade most surfaces, especially between moving parts. Air and oil filters will become blocked requiring frequent replacement. Rail transport is less vulnerable, with disruptions mainly caused by reduction in visibility.

Marine transport can also be impacted by volcanic ash. Ash fall will block air and oil filters and abrade any moving parts if ingested into engines. Navigation will be impacted by a reduction in visibility during ash fall. Vesiculated ash (pumice and scoria) will float on the water surface in 'pumice rafts' which can clog water intakes quickly, leading to over heating of machinery.

Communications

Telecommunication and broadcast networks can be affected by volcanic ash in the following ways: attenuation and reduction of signal strength; damage to equipment; and overloading of network through user demand. Signal attenuation due to volcanic ash is not well documented; however, there have been reports of disrupted communications following the 1969 Surtsey eruption and 1991 Mount Pinatubo eruption. Research by the New Zealand-based Auckland Engineering Lifelines Group determined theoretically that impacts on telecommunications signals from ash would be limited to low frequency services such as satellite communication. Signal interference may also be caused by lightning, as this is frequently generated within volcanic eruption plumes.

Telecommunication equipment may become damaged due to direct ash fall. Most modern equipment requires constant cooling from air conditioning units. These are susceptible to blockage by ash which reduces their cooling efficiency. Heavy ash falls may cause telecommunication lines, masts, cables, aerials, antennae dishes and towers to collapse due to ash loading. Moist ash may also cause accelerated corrosion of metal components.

Reports from recent eruptions suggest that the largest disruption to communication networks is overloading due to high user demand. This is common of many natural disasters.

Computers

Computers may be impacted by volcanic ash, with their functionality and usability decreasing during ashfall, but it is unlikely they will completely fail. The most vulnerable components are the mechanical components, such as cooling fans, cd drives, keyboard, mice and touch pads. These components can become jammed with fine grained ash causing them to cease working; however, most can be restored to working order by cleaning with compressed air. Moist ash may cause electrical short circuits within desktop computers; however, will not affect laptop computers.

Buildings and Structures

Damage to buildings and structures can range from complete or partial roof collapse to less catastrophic damage of exterior and internal materials. Impacts depend on the thickness of ash, whether it is wet or dry, the roof and building design and how much ash gets inside a building. The specific weight of ash can vary significantly and rain can increase this by 50-100%. Problems associated with ash loading are similar to that of snow; however, ash is more severe as 1) the load from ash is generally much greater, 2) ash does not melt and 3) ash can clog and damage gutters, especially after rain fall. Impacts for ash loading depend on building design and construction, including roof slope, construction materials, roof span and support system, and age and maintenance of the building. Generally flat roofs are more susceptible to damage and collapse than steeply pitched roofs. Roofs made of smooth materials (sheet metal or glass) are more likely to shed ash than roofs made with rough materials (thatch, asphalt or wood shingles). Roof collapse can lead to widespread injuries and deaths and property damage. For example, the collapse of roofs from ash during the 15 June 1991 Mount Pinatubo eruption killed about 300 people.

Human and Animal Health

Ash particles of less than 10 μm diameter suspended in the air are known to be inhalable, and people exposed to ash falls have experienced respiratory discomfort, breathing difficulty, eye and skin irritation, and nose and throat symptoms. Most of these effects are short-term and are not considered to pose a significant health risk to those without pre-existing respiratory conditions. The health effects of volcanic ash depend on the grain size, mineralogical composition and chemical coatings on the surface of the ash particles. Additional factors related to potential respiratory symptoms are the frequency and duration of exposure, the concentration of ash in the air and the respirable ash fraction; the proportion of ash with less than 10 μm diameter, known as PM_{10}. The social context may also be important.

Chronic health effects from volcanic ash fall are possible, as exposure to free crystalline silica is known to cause silicosis. Minerals associated with this include quartz, cristobalite and tridymite, which may all be present in volcanic ash. These minerals are described as 'free' silica as the SiO_2 is not attached to another element to create a new mineral. However, magmas containing less than 58% SiO_2 are thought to be unlikely to contain crystalline silica.

The exposure levels to free crystalline silica in the ash are commonly used to characterise the risk of silicosis in occupational studies (for people who work in mining, construction and other industries,) because it is classified as a human carcinogen by the International Agency for Research on Cancer. Guideline values have been created for exposure, but with unclear rationale; UK guidelines for particulates in air (PM10) are 50 μg/m^3 and USA guidelines for exposure to crystalline silica are 50 μg/m^3. It is thought that the guidelines on exposure levels could be exceeded for short periods of time without significant health effects on the general population.

There have been no documented cases of silicosis developed from exposure to volcanic ash. However, long-term studies necessary to evaluate these effects are lacking.

Ingesting ash may be harmful to livestock, causing abrasion of the teeth, and in cases of high fluorine content, fluorine poisoning (toxic at levels of >100 μg/g) for grazing animals. It is known from the 1783 eruption of Laki in Iceland that fluorine poisoning occurred in humans and livestock as a result of the chemistry of the ash and gas, which contained high levels of Hydrogen Fluoride. Following the 1995/96 Mount Ruapehu eruptions in New Zealand, two thousand ewes and lambs died after being affected by fluorosis while grazing on land with only 1–3 mm of ash fall. Ash ingestion may also cause gastrointestinal blockages. Sheep that ingested ash from the 1991 Mount Hudson volcanic eruption in Chile, suffered from diarrhoea and weakness. The added weight of ash in the wool led to fatigue and sheep could not stand up.

Environment and Agriculture

Volcanic ash can have a detrimental impact on the environment which can be difficult to predict due to the large variety of environmental conditions that exist within the ash fall zone. Natural waterways can be impacted in the same way as urban water supply networks. Ash will increase water turbidity which can reduce the amount of light reaching lower depths, which can inhibit growth of submerged aquatic plants and consequently affect species which are dependent on them such as fish and shellfish. High turbidity can also affect the ability of fish gills to absorb dissolved oxygen. Acidification will also occur, which will reduce the pH of the water and impact the fauna and flora living in the environment. Fluoride contamination will occur if the ash contains high concentrations of fluoride.

Ash accumulation will also affect pasture, plants and trees which are part of the horticulture and agriculture industries. Thin ash falls (<20 mm) may put livestock off eating, and can inhibit transpiration and photosynthesis and alter growth. There may be an increase in pasture production due to a mulching effect and slight fertilizing effect, such as occurred following the 1980 Mount St. Helens and 1995/96 Mt Ruapehu eruptions. Heavier falls will completely bury pastures and soil leading to death of pasture and sterilization of the soil due to oxygen deprivation. Plant survival is dependent on ash thickness, ash chemistry, compaction of ash, amount of rainfall, duration of burial and the length of plant stalks at the time of ash fall. The acidic nature of ash will lead to elevated soil sulfur levels and lowered soil pH, which can reduce the availability of essential minerals and alter the soil's characteristics so that crops and plants will not survive. Ash will also impact upon arable crops, such as fruit, vegetables and grain. Ash can burn plant and crop tissue reducing quality, contaminate crops during harvest and damage plants from ash loading.

Young forests (trees <2 years old) are most at risk from ash falls and are likely to be destroyed by ash deposits >100 mm. Ash fall is unlikely to kill mature trees, but ash loading may break large branches during heavy ash falls (>500 mm). Defoliation of trees may also occur, especially if there is a coarse ash component within the ash fall.

Land rehabilitation after ash fall may be possible depending on the ash deposit thickness. Rehabilitation treatment may include: direct seeding of deposit; mixing of deposit with buried soil; scraping of ash deposit from land surface; and application of new topsoil over the ash deposit.

Interdependence

Critical infrastructure and infrastructure services are vital to the functionality of modern society, to provide: medical care, policing, emergency services, and lifelines such as water, wastewater, and power and transportation links. Often critical facilities themselves are dependent on such lifelines for operability, which makes them vulnerable to both direct impacts from a hazard event and indirect effects from lifeline disruption.

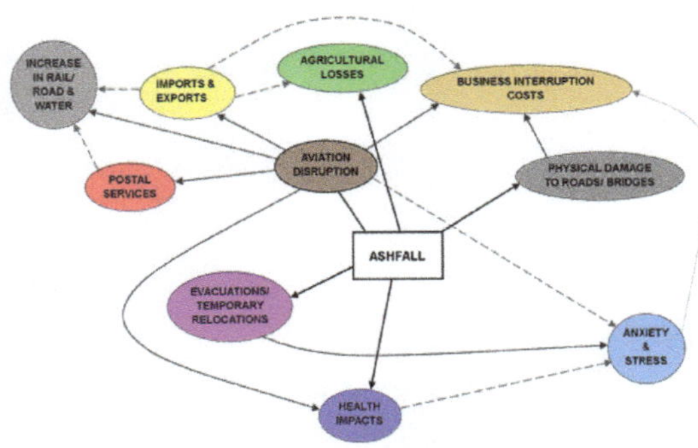

Interdependency of volcanic ashfall impacts from the Eyjafjallajökull 2010 eruptions.

The impacts on lifelines may also be inter-dependent. The vulnerability of each lifeline may depend on: the type of hazard, the spatial density of its critical linkages, the dependency on critical linkages, susceptibility to damage and speed of service restoration, state of repair or age, and institutional characteristics or ownership.

The 2010 eruption of Eyjafjallajokull in Iceland highlighted the impacts of volcanic ash fall in modern society and our dependence on the functionality of infrastructure services. During this event the airline industry suffered business interruption losses of €1.5-2.5 billion from the closure of European airspace for six days in April 2010 and subsequent closures into May 2010. Ash fall from this event is also known to have caused local crop losses in agricultural industries, losses in the tourism industry, destruction of roads and bridges in Iceland (in combination with glacial melt water), and costs associated with emergency response and clean-up. However, across Europe there were further losses associated with travel disruption, the insurance industry, the postal service, and imports and exports across Europe and worldwide. These consequences demonstrate the interdependency and diversity of impacts from a single event.

Preparedness, Mitigation and Management

Two management methods during the 2014 eruptions of Kelud: sweeping (top) and spraying with water (bottom)

Preparedness for ashfalls should involve sealing buildings, protecting infrastructure and homes, and storing sufficient supplies of food and water to last until the ash fall is over and clean-up can begin. Dust masks can be worn to reduce inhalation of ash and mitigate against any respiratory health affects. Goggles can be worn to protect against eye irritation.

The International Volcanic Ashfall Impacts Working Group of IAVCEI maintains a regularly updated database of impacts and mitigations strategies at http://volcanoes.usgs.gov/ash

At home, staying informed about volcanic activity, and having contingency plans in place for alternative shelter locations, constitutes good preparedness for an ash fall event. This can prevent some impacts associated with ash fall, reduce the effects, and increase the human capacity to cope with such events. A few items such as a flashlight, plastic sheeting to protect electronic equipment from ash ingress, and battery operated radios, are extremely useful during ash fall events.

The protection of infrastructure must also be considered within emergency preparedness. Critical facilities that need to remain operable should be identified, and all others should be shut down to reduce damage. It is also important to keep ash out of buildings, machinery and lifeline networks (in particular water and wastewater systems,) to prevent some of the damage caused by ash particles. Windows and doors should be closed and shuttered if possible, to prevent ingress of ash into buildings.

Communication plans should be made beforehand to inform of mitigation actions being undertaken. Spare parts and back-up systems should be in place prior to ash fall events to reduce service disruption and return functionality as quickly as possible. Good preparedness also includes the identification of ash disposal sites, before ash fall occurs, to avoid further movement of ash and to aid clean-up. Protective equipment such as eye protection and dust masks should be deployed for clean-up teams in advance of ash fall events.

Some effective techniques for the management of ash have been developed including cleaning methods and cleaning apparatus, and actions to mitigate or limit damage. The latter include covering of openings such as: air and water intakes, aircraft engines and windows during ash fall events. Roads may be closed to allow clean-up of ash falls, or speed restrictions may be put in place, in order to prevent motorists from developing motor problems and becoming stranded following an ash fall. To prevent further effects on underground water systems or waste water networks, drains and culverts should be unblocked and ash prevented from entering the system. Ash can be moistened (but not saturated) by sprinkling with water, to prevent remobilisation of ash and to aid clean-up. Prioritisation of clean-up operations for critical facilities and coordination of clean-up efforts also constitute good management practice.

Volcanic Ash Soils

Volcanic ash's primary use is that of a soil enricher. Once the minerals in ash are washed into the soil by rain or other natural processes, it mixes with the soil to create an Andisol layer. This layer is highly rich in nutrients and is very good for agricultural use; the presence of lush forests on volcanic islands is often as a result of trees growing and flourishing in the Phosphorus and Nitrogen-rich andisol.

Volcanic Arc

A volcanic arc is a chain of volcanoes formed above a subducting plate, positioned in an arc shape as seen from above. Offshore volcanoes form islands, resulting in a volcanic island arc. Generally they result from the subduction of an oceanic tectonic plate under another tectonic plate, and often parallel an oceanic trench. The oceanic plate is saturated with water, and volatiles such as water drastically lower the melting point of the mantle. As the oceanic plate is subducted, it is subjected to greater and greater pressures with increasing depth. This pressure squeezes water out of the plate and introduces it to the mantle. Here the mantle melts and forms magma at depth under the overriding plate. The magma ascends to form an arc of volcanoes parallel to the subduction zone.

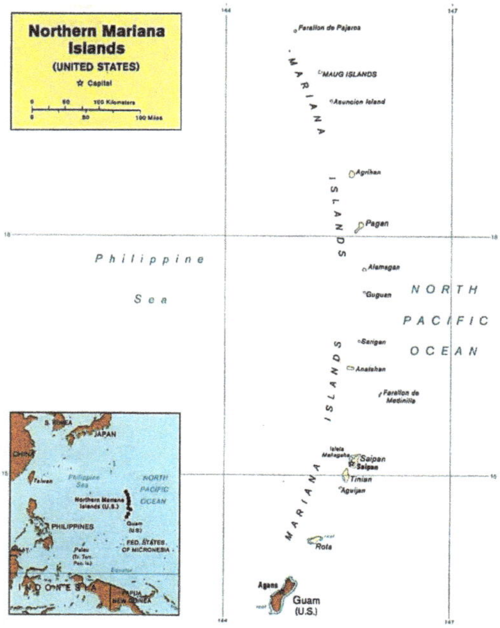

Mariana Islands, an oceanic island arc

These should not be confused with hotspot volcanic chains, where volcanoes often form one after another in the middle of a tectonic plate, as the plate moves over the hotspot, and so the volcanoes progress in age from one end of the chain to the other. The Hawaiian Islands form a typical hotspot chain; the older islands (tens of millions of years) to the northwest are smaller and more lush than the recently created (400,000 years ago) Hawaii island itself, which is more rocky. Hot spot volcanoes are also known as "Intra-plate" volcanoes and the islands they create are known as Volcanic Ocean Islands. Volcanic arcs do not generally exhibit such a simple age pattern.

There are two types of volcanic arcs:

- oceanic arcs form when oceanic crust subducts beneath other oceanic crust on an adjacent plate, creating a volcanic island arc. (Not all island arcs are volcanic island arcs.)
- continental arcs form when oceanic crust subducts beneath continental crust on an adjacent plate, creating an arc-shaped mountain belt. (Not all mountain belts are formed this way.)

In some situations, a single subduction zone may show both aspects along its length, as part of a plate subducts beneath a continent and part beneath adjacent oceanic crust.

(The term "volcanic arc" is often confused with the term "island arc". The former is a series of volcanoes, but not necessarily offshore. The latter is a series of islands, not necessarily composed solely of volcanoes.)

Volcanoes are present in almost any mountain belt, but this does not make it a **volcanic** arc. Often there are isolated, but impressively huge volcanoes in a mountain belt. For instance, Vesuvius and the Etna volcanoes in Italy are part of separate but different kinds of mountainous volcanic ensembles.

The active front of a volcanic arc is the belt where volcanism develops at a given time. Active fronts may move over time (millions of years) changing their distance from the oceanic trench as well as their width.

Petrology

In the subduction zone, loss of water from the subducted slab induces partial melting of the overriding mantle and generates low-density, calc-alkaline magma that buoyantly rises to intrude and be extruded through the lithosphere of the overriding plate. This loss of water is due to the destabilization of the mineral chlorite at approximately 40–60 km depth. This is the reason for island arc volcanism at consistent distances from the subducting slab: because the temperature-pressure conditions for flux-melting volcanism due to chlorite destabilization will always occur at the same depth, the distance from the trench to the arc volcanoes is determined only by the dip angle of the subducting slab.

On the subducting side of the island arc is a deep and narrow oceanic trench, which is the trace at the Earth's surface of the boundary between the downgoing and overriding plates. This trench is created by the gravitational pull of the relatively dense subducting plate pulling the leading edge of the plate downward. Multiple earthquakes occur along this subduction boundary with the seismic hypocenters located at increasing depth under the island arc: these quakes define the Wadati-Benioff zones. The volcanic arc forms when the subducting plate reaches a depth of about 100 km.

Ocean basins that are being reduced by subduction are called 'remnant oceans' as they will slowly be shrunken out of existence and crushed in the subsequent orogenic collision. This process has happened over and over in the geologic history of the Earth.

In the rock record, volcanic arcs can be seen as the volcanic rocks themselves, but because volcanic rock is easily weathered and eroded, it is more typical that they are seen as plutonic rocks, the rocks that formed underneath the arc (e.g. the Sierra Nevada batholith), or in the sedimentary record as lithic sandstones.

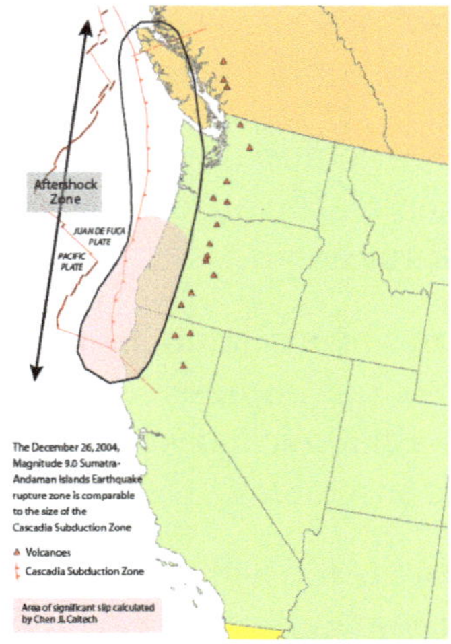

Cascade Volcanic Arc, a continental volcanic arc

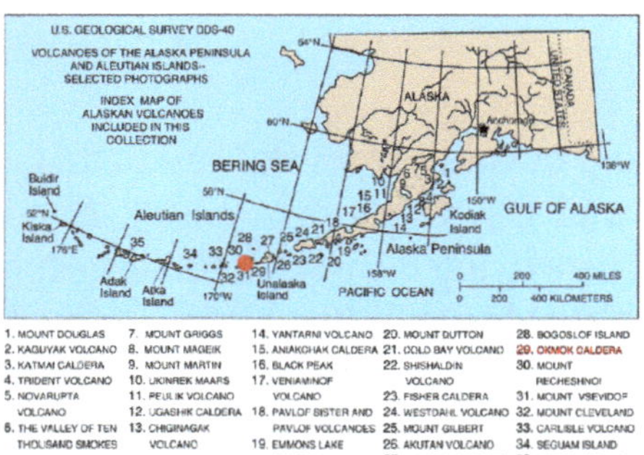

The Aleutian Arc, with both oceanic and continental parts.

Examples

Two classic examples of oceanic island arcs are the Mariana Islands in the western Pacific Ocean and the Lesser Antilles in the western Atlantic Ocean. The Cascade Volcanic Arc in western North America and the Andes along the western edge of South America are examples of continental volcanic arcs. The best examples of volcanic arcs with both sets of characteristics are in the North Pacific, with the Aleutian Arc consisting of the Aleutian Islands and their extension the Aleutian Range on the Alaska Peninsula, and the Kuril-Kamchatka Arc comprising the Kuril Islands and southern Kamchatka Peninsula.

Continental Arcs

- Cascade Volcanic Arc
- Alaska Peninsula and Aleutian Range
- Kamchatka
- Andes
    - North Volcanic Zone

    - Central Volcanic Zone
    - South Volcanic Zone
    - Austral Volcanic Zone
- Central America Volcanic Arc

Island Arcs

- Aleutian Islands
- Kuril Islands
- Northeastern Japan arc
- Japanese Archipelago including the Ryukyu Islands
- Izu-Bonin-Mariana Arc:
    - Izu Islands
    - Bonin Islands
    - Mariana Islands
- Luzon Volcanic Arc
- Philippines
- Tonga and Kermadec Islands
- Andaman and Nicobar Islands
- Mentawai Islands
- Sunda Arc
    - Lesser Sunda Islands
- Tanimbar and Kai Islands
- Solomon Islands
- South Aegean Volcanic Arc
- Egean, or Hellenic arc
- Lesser Antilles, including the Leeward Antilles
- Scotia Arc:
    - South Sandwich Islands
- Mascarene Islands

Ancient Island Arcs

- Insular Islands
- Intermontane Islands
- Sakhalin island arc

References

- Decker, Robert Wayne; Decker, Barbara (1991). Mountains of Fire: The Nature of Volcanoes. Cambridge University Press. p. 7. ISBN 0-521-31290-6. Retrieved August 16, 2012.
- Volcanoes in human history: the far-reaching effects of major eruptions. Jelle Zeilinga de Boer, Donald Theodore Sanders (2002). Princeton University Press. p. 155. ISBN 0-691-05081-3
- Benton M J (2005). When Life Nearly Died: The Greatest Mass Extinction of All Time. Thames & Hudson. ISBN 978-0-500-28573-2.
- McBirney AR (1996). "The Skaergaard intrusion". In Cawthorn RG. Layered intrusions. Developments in petrology. 15. pp. 147–180. ISBN 9780080535401.
- Bindschadler, D. L. (1995). "Magellan: A new view of Venus' geology and geophysics". Reviews of Geophysics. 33: 459. doi:10.1029/95RG00281. Retrieved 28 September 2015.
- "Types and Processes Gallery: Lava Flows". Global Volcanism Program. Smithsonian Institution. 2013. Retrieved 1 December 2015.
- Popp, B. N. (September 2007). "Igneous Rocks" (PDF). Geology Course Lecture in PowerPoint. School of Ocean and Earth Science and Technology, University of Hawai'i at Manoa. Retrieved 16 October 2015.
- NSTA Press / Archive.Org (2007). "Earthquakes, Volcanoes, and Tsunamis" (PDF). Resources for Environmental Literacy. Archived from the original (PDF) on July 14, 2014. Retrieved April 22, 2014.
- International Volcanic Health Hazard Network. "International Volcanic Health Hazard Network". Retrieved 30 November 2011.
- Chesner, C.A.; Rose, J.A.; Deino, W.I.; Drake, R.; Westgate, A. (March 1991). "Eruptive History of Earth's Largest Quaternary caldera (Toba, Indonesia) Clarified" (PDF). Geology. 19 (3): 200–203. Bibcode:1991Geo....19..200C. doi:10.1130/0091-7613(1991)019<0200:EHOESL>2.3.CO;2. Retrieved January 20, 2010.
- Jaggard, Victoria (February 5, 2010). ""Super Earth" May Really Be New Planet Type: Super-Io". National Geographic web site daily news. National Geographic Society. Retrieved March 11, 2010.

3

Classification of Volcanoes

Volcanoes can best be understood in confluence with the major topics listed in the following chapter. Some of the classifications of the volcanoes enumerated in this chapter are stratovolcanoes, shield volcanoes, mud volcanoes etc. Whereas stratovolcanoes erupt periodically, shield volcanoes are entirely built with magma fluid, and once shield volcanoes erupt, the lava travels farther than the lava erupted by stratovolcanoes.

Stratovolcano

A stratovolcano, also known as a composite volcano, is a conical volcano built up by many layers (strata) of hardened lava, tephra, pumice, and volcanic ash. Unlike shield volcanoes, stratovolcanoes are characterized by a steep profile and periodic explosive eruptions and effusive eruptions, although some have collapsed craters called calderas. The lava flowing from stratovolcanoes typically cools and hardens before spreading far due to high viscosity. The magma forming this lava is often felsic, having high-to-intermediate levels of silica (as in rhyolite, dacite, or andesite), with lesser amounts of less-viscous mafic magma. Extensive felsic lava flows are uncommon, but have travelled as far as 15 km (9.3 mi).

Mount Vesuvius near Naples, Italy, erupted in 79 AD. The last eruption of this stratovolcano occurred in March 1944.

Stratovolcanoes are sometimes called "composite volcanoes" because of their composite layered structure built up from sequential outpourings of eruptive materials. They are among the most common types of volcanoes, in contrast to the less common shield volcanoes. Two famous stratovolcanoes are Krakatoa, best known for its catastrophic eruption in 1883 and Vesuvius, famous for its destruction of the towns Pompeii and Herculaneum in 79 CE. Both eruptions claimed thousands of lives. In modern times, Mount Saint Helens and Mount Pinatubo have erupted catastrophically.

Existence of stratovolcanoes has not been proved on other terrestrial bodies of the solar system with one exception. Their existence was suggested for some isolated massifs on Mars, e.g., Zephyria Tholus.

Creation

Cross-section of subduction zone and associated stratovolcano

Stratovolcanoes are common at subduction zones, forming chains along plate tectonic boundaries where oceanic crust is drawn under continental crust (continental arc volcanism, e.g. Cascade Range, central Andes) or another oceanic plate (island arc volcanism, e.g. Japan, Aleutian Islands). The magma forming stratovolcanoes rises when water trapped both in hydrated minerals and in the porous basalt rock of the upper oceanic crust is released into mantle rock of the asthenosphere above the sinking oceanic slab. The release of water from hydrated minerals is termed "dewatering", and occurs at specific pressures and temperatures for each mineral, as the plate descends to greater depths. The water freed from the rock lowers the melting point of the overlying mantle rock, which then undergoes partial melting and rises due to its lighter density relative to the surrounding mantle rock, and pools temporarily at the base of the lithosphere. The magma then rises through the crust, incorporating silica-rich crustal rock, leading to a final intermediate composition. When the magma nears the top surface, it pools in a magma chamber under or within the volcano.

There, the relatively low pressure allows water and other volatiles (mainly CO_2, SO_2, Cl_2, and H_2O) dissolved in the magma to escape from solution, as occurs when a bottle of carbonated water is opened, releasing CO_2. Once a critical volume of magma and gas accumulates, the obstacle (rock blockage) of the volcanic cone is overcome, leading to a sudden explosive eruption.

Hazards

Mount Etna on the Island of Sicily

In recorded history, explosive eruptions at subduction zone (convergent-boundary) volcanoes have posed the greatest hazard to civilizations. Subduction-zone stratovolcanoes, such as Mount St. Helens, Mount Etna, and Mount Pinatubo, typically erupt with explosive force: the magma is too stiff to allow easy escape of volcanic gases. As a consequence, the tremendous internal pressures of the trapped volcanic gases remain in the pasty magma. Following the breaching of the magma chamber, the magma degasses explosively. The gases and water gush out with speed and force.

Since 1600 CE, nearly 300,000 people have been killed by volcanic eruptions. Most deaths were caused by pyroclastic flows and mudflows, deadly hazards that often accompany explosive eruptions of subduction-zone stratovolcanoes. Pyroclastic flows are fast-moving, avalanche-like, ground-hugging, incandescent mixtures of hot volcanic debris, ash, and gases that can travel at speeds in excess of 160 km/h (100 mph). Around 30,000 people were killed by pyroclastic flows during the 1902 eruption of Mont Pelée on the island of Martinique in the Caribbean. In March–April 1982, three explosive eruptions of El Chichón Volcano in the State of Chiapas, southeastern Mexico, caused the worst volcanic disaster in that country's history. Villages within 8 km (5 mi) of the volcano were destroyed by pyroclastic flows, killing more than 2,000 people.

Mount Unzen in southwestern Japan

Two Decade Volcanoes that erupted in 1991 provide examples of stratovolcano hazards. On June 15, Mount Pinatubo spewed ash 40 km (25 mi) into the air and produced huge pyroclastic flows and mudflows that devastated a large area around the volcano. Pinatubo, located 90 km (56 mi) from Manila, had been dormant for 600 years before the 1991 eruption, which ranks as one of the largest eruptions in the 20th century. Also in 1991, Japan's Unzen Volcano, located on the island of Kyushu about 40 km (25 mi) east of Nagasaki, awakened from its 200-year slumber to produce a new lava dome at its summit. Beginning in June, repeated collapse of this erupting dome generated ash flows that swept down the mountain's slopes at speeds as high as 200 km/h (120 mph). Unzen is one of more than 75 active volcanoes in Japan; an eruption in 1792 killed more than 15,000 people—the worst volcanic disaster in the country's history.

The AD 79 Plinian eruption of Mount Vesuvius, a stratovolcano looming adjacent to Naples, completely covered the cities of Pompeii and Herculaneum with pyroclastic surge deposits. The death toll ranged between 10,000 and 25,000. Mount Vesuvius is recognized as one of the most dangerous volcanoes, jointly because of its potential for powerful explosive eruptions and the high population density of the area (totaling about 3 million people) around its perimeter.

Ash

Snowlike blanket of Mount Pinatubo's ashfall deposits of June 15, 1991.

Apart from possibly affecting the climate, volcanic clouds from explosive eruptions also pose a serious hazard to aviation safety. For example, during the 1982 eruption of Galunggung in Java, British Airways Flight 9 flew into the ash cloud, suffering temporary engine failure and structural damage. During the past two decades, more than 60 airplanes, mostly commercial jetliners, have been damaged by in-flight encounters with volcanic ash. Some of these encounters have resulted in the power loss of all engines, necessitating emergency landings. Luckily, to date no crashes have happened because of jet aircraft flying into volcanic ash.

Ash fall is a threat to health when inhaled, and is also a threat to property with enough accumulation. An accumulation of 30 cm (12 in) is sufficient to cause most buildings to collapse. Dense clouds of hot volcanic ash, caused by the collapse of an eruptive column or by being laterally expelled from the partial collapse of a volcanic edifice or lava dome during an explosive eruption, can produce devastating pyroclastic flows which can wipe out everything in their paths.

Lava

Lava flows from stratovolcanoes are generally not a significant threat to people because the highly viscous lava moves slowly enough for people to move out of the path of flow. The lava flows are more of a threat to property.

Mayon Volcano produced lava flows during its eruption on December 29, 2009.

However, not all stratovolcanoes erupt viscous, blocky lava. Mount Nyiragongo is very dangerous because its magma has an unusually low silica content, making it quite fluid. Fluid lavas are typically associated with the formation of broad shield volcanoes such as those of Hawaii, but Nyiragongo has very steep slopes down which lava can flow at up to 100 km/h (62 mph). Lava flows could melt down ice and glaciers that accumulated on the volcano's crater and upper slopes, generating massive lahar flows. Rarely, generally fluid lava could also generate a massive lava fountain, while lava of thicker viscosity can solidify within the vent, creating a block which can result in highly explosive eruptions.

Volcanic Bombs

Volcanic bombs are extrusive igneous rocks ranging from the size of a book to small cars, that are explosively ejected from stratovolcanoes during their climactic eruptive phases. These "bombs" can travel over 20 km (15 mi) away from the volcano, and present a risk to buildings and living things while traveling at very high speeds (hundreds of kilometers/miles per hour) through the air. Most bombs do not themselves explode on impact, but rather carry enough force so as to have destructive effects as if they exploded.

Mudflows

Mudflows (also called debris flows or lahars, a Javanese term for volcanic mudflows) are mixtures of volcanic debris and water. The water usually comes from two sources: rainfall or the melting of snow and ice by hot volcanic debris, such as lava. Depending on the proportion of water to volcanic material, mudflows can range from fast-flowing, soupy floods to thick, gooey flows that have the consistency of wet concrete. As mudflows sweep down the steep sides of composite volcanoes, they have the strength and speed to flatten or bury everything in their paths. Hot ash, lava and pyroclastic flows from the 1985 eruption of the Nevado del Ruiz Volcano in Colombia melted snow and ice atop the 5,390-m-high Andean peak; the ensuing mudflows buried the nearby city of Armero, killing 25,000 people.

Climatic Effects

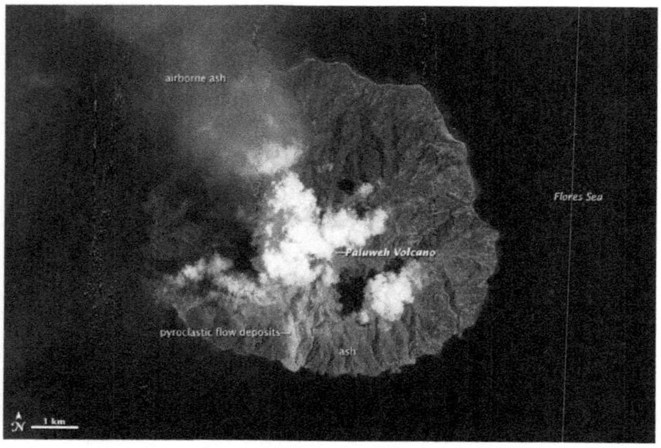

Paluweh eruption as seen from space

As per the above examples, while the Unzen eruptions have caused deaths and considerable local damage in the historic past, the impact of the June 1991 eruption of Mount Pinatubo was global. Slightly cooler-than-usual temperatures were recorded worldwide, and brilliant sunsets and intense sunrises were attributed to the particulates, this eruption lofted high into the stratosphere. The aerosol that formed from the sulfur dioxide (SO_2), carbon dioxide (CO_2) and other gasses dispersed around the world. The SO_2 mass in this cloud—about 22 million tons—combined with water (both of volcanic and stratospheric origin) formed droplets of sulfuric acid, blocking a portion of the sunlight from reaching the troposphere and ground. The cooling in some regions is thought to have been as much as 0.5 °C. An eruption the size of Mount Pinatubo tends to affect the weather for a few years; the material injected into the stratosphere gradually drops into the troposphere, where it is washed away by rain and cloud precipitation.

A similar, but extraordinarily more powerful phenomenon occurred in the cataclysmic April 1815 eruption of Mount Tambora on Sumbawa Island in Indonesia. The Mount Tambora eruption is recognized as the most powerful eruption in recorded history. Its volcanic cloud lowered global temperatures by as much as 3.5 °C. In the year following the eruption, most of the Northern Hemisphere experienced sharply cooler temperatures during the summer. In parts of Europe, Asia and North America, 1816 was known as the "Year Without a Summer", which caused a considerable agricultural distress and a brief but bitter famine.

Shield Volcano

A shield volcano is a type of volcano usually built almost entirely of fluid magma flows. They are named for their large size and low profile, resembling a warrior's shield lying on the ground. This is caused by the highly fluid lava they erupt, which travels farther than lava erupted from stratovolcanoes. This results in the steady accumulation of broad sheets of lava, building up the shield volcano's distinctive form. The shape of shield volcanoes is due to the low-viscosity magma of their mafic lava.

Mauna Kea, Hawai□i, a shield volcano on the Big Island of Hawaii

An Ancient Greek warrior's shield–its circular shape and gently sloping surface, with a central raised area, is a shape shared by many shield volcanoes.

Etymology

Shield volcanoes are built by effusive eruptions, which flow out in all directions to create a shield like that of a warrior. The word "shield" has a long history, and is derived from the Old English *scield* or *scild*, which is in turn taken from the Proto-Germanic **skelduz*, and related to the Gothic *skildus*, meaning "to divide, split, or separate". Shield volcano itself is taken from the German term *Schildvulkan*.

Geology

Structure

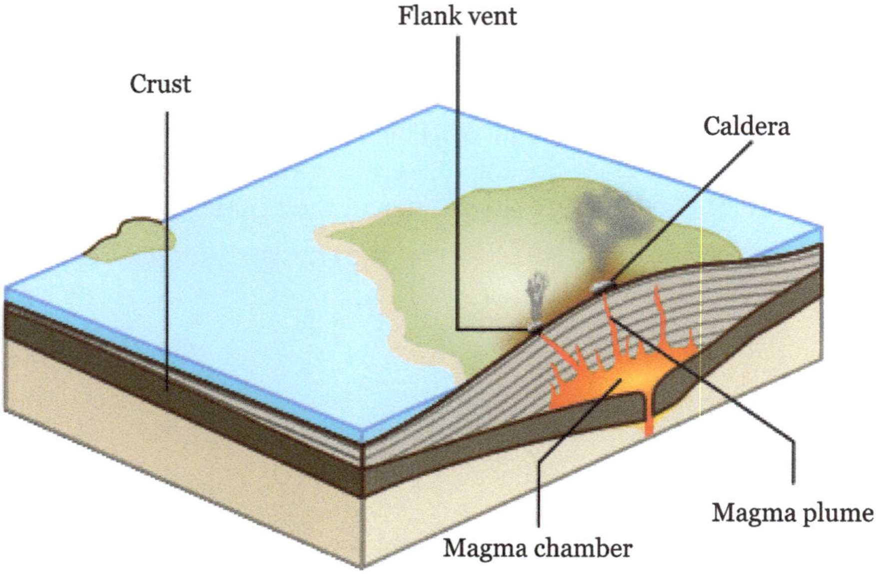

Diagram of the common structural features of a shield volcano.

Shield volcanoes are distinguished from the three other major volcanic archetypes—stratovolcanoes, lava domes, and cinder cones—by their structural form, a consequence of their unique magmatic composition. Of these four forms shield volcanoes erupt the least viscous lavas: where stratovolcanoes and especially lava domes are the product of highly immotile flows and cinder cones are constructed by explosively eruptive tephra, shield volcanoes are the product of gentle effusive eruptions of highly fluid lavas that produce, over time, a broad, gently sloped eponymous "shield". Although the term is generally ascribed to basaltic shields it has also at times been appended to rarer scutiform volcanoes of differing magmatic composition—principally pyroclastic shields, formed by the accumulation of fragmental material from particularly powerful explosive eruptions, and rarer felsic lava shields formed by unusually fluid felsic magmas. Examples of pyroclastic shields include Billy Mitchell volcano in Papua New Guinea and the Purico Complex in Chile; an example of a felsic shield is the Big Obsidian Flow in Oregon. Shield volcanoes are also related in origination to vast lava plateaus and flood basalts present in various parts of the world, generalized eruptive features which occur along linear fissure vents and are distinguished from shield volcanoes proper by the lack of an identifiable primary eruptive center.

Active shield volcanoes experience near-continuous eruptive activity over extremely long periods of time, resulting in the gradual build-up of edifices that can reach extremely large dimensions. With the exclusion of flood basalts mature shields are the largest volcanic features on Earth: the summit of the largest subaerial volcano in the world, Mauna Loa, lies 4,169 m (13,678 ft) above sea level, and the volcano, over 60 mi (97 km) wide at its base, is estimated to contain about 80,000 km^3 (19,000 cu mi) of basalt. The mass of the volcano is so great that it has slumped the crust beneath it a further 8 km (5 mi); accounting for this subsidence and for the height of the volcano above the sea floor the "true" height of Mauna Loa from the start of its eruptive history is about 17,170 m (56,000 ft). Mount Everest, by comparison, is 8,848 m (29,029 ft) in height. In September 2013 a team led by the University of Houston's William Sager announced the singular origination of Tamu Massif, an enormous extinct submarine shield volcano of previously unknown origin which, approximately 450 by 650 km (280 by 400 mi) in area, dwarfs all previously known volcanoes on the planet. The research has not yet been confirmed.

Shield volcanoes feature a gentle (usually 2 to 3°) slope that gradually steepens with elevation (reaching approximately 10°) before eventually flattening near the summit, forming an overall upwardly convex shape. In height they are typically about one twentieth their width. Although the general form of a "typical" shield volcano varies little worldwide regional differences exist in their size and morphological characteristics. Typical shield volcanoes present in California and Oregon measure 3 to 4 mi (5 to 6 km) in diameter and 1,500 to 2,000 ft (460 to 610 m) in height; shield volcanoes in the central Mexican Michoacán-Guanajuato volcanic field, by comparison, average 340 m (1,115 ft) in height and 4,100 m (13,451 ft) in width, with an average slope angle of 9.4° and an average volume of 1.7 km^3 (0 cu mi).

Rift zones are a prevalent feature on shield volcanoes that is rare on other volcanic types. The large, decentralized shape of Hawaiian volcanoes as compared to their smaller, symmetrical Icelandic cousins can be attributed to rift eruptions. Fissure venting is common in Hawai□i; most Hawaiian eruptions begin with a so-called "wall of fire" along a major fissure line before centralizing to a small number of points. This accounts for their asymmetrical shape, whereas Icelandic volcanoes follow a pattern of central eruptions dominated by summit calderas, causing the lava to be more evenly distributed or symmetrical.

Eruptive Characteristics

Most of what is currently known about shield volcanic eruptive character has been gleaned from studies done on the volcanoes of Hawai□i island, by far the most intensively studied of all shields due to their scientific accessibility; the island lends its name to the slow-moving, effusive eruptions typical of shield volcanism, known as Hawaiian eruptions. These eruptions, the calmest of volcanic events, are characterized by the effusive emission of highly fluid basaltic lavas with low gaseous content. These lavas travel a far greater distance than those of other eruptive types before solidifying, forming extremely wide but relatively thin magmatic sheets often less than 1 m (3.3 ft) thick. Low volumes of such lavas layered over long periods of time are what slowly constructs the characteristically low, broad profile of a mature shield volcano.

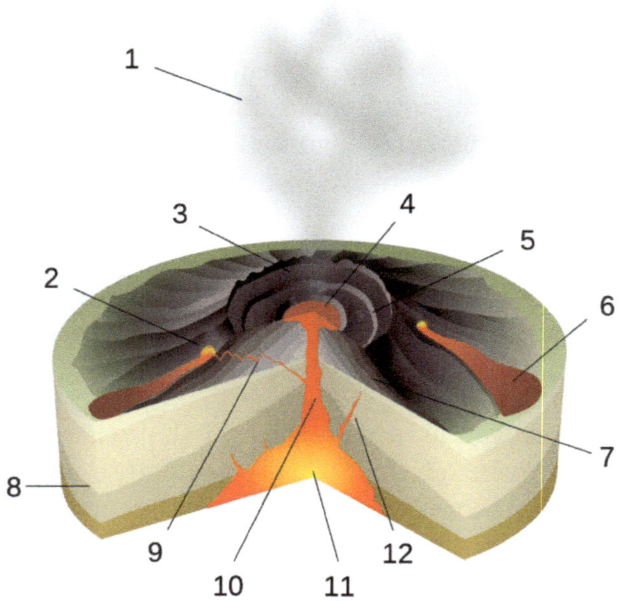

Diagram of a Hawaiian eruption. (key: 1. Ash plume 2. Lava fountain 3. Crater 4. Lava lake 5. Fumaroles 6. Lava flow 7. Layers of lava and ash 8. Stratum 9. Sill 10. Magma conduit 11. Magma chamber 12. Dike) Click for larger version.

Also unlike other eruptive types, Hawaiian eruptions often occur at decentralized fissure vents, beginning with large "curtains of fire" that quickly die down and concentrate at specific locations on the volcano's rift zones. Central-vent eruptions, meanwhile, often take the form of large lava fountains (both continuous and sporadic), which can reach heights of hundreds of meters or more. The particles from lava fountains usually cool in the air before hitting the ground, resulting in the accumulation of cindery scoria fragments; however, when the air is especially thick with clasts, they cannot cool off fast enough due to the surrounding heat, and hit the ground still hot, accumulating into spatter cones. If eruptive rates are high enough, they may even form splatter-fed lava flows. Hawaiian eruptions are often extremely long lived; Pu'u'Ō'ō, a cinder cone of Kilauea, has been erupting continuously since 1983.

Flows from Hawaiian eruptions can be divided into two types by their structural characteristics: pāhoehoe lava which is relatively smooth and flows with a ropey texture, and lava flows which are denser, more viscous (and thus slower moving) and blockier. These lava flows can be anywhere between 2 and 20 m (7 and 66 ft) thick. Lava flows move through pressure—the partially solidi-fied front of the flow steepens due to the mass of flowing lava behind it until it breaks off, after which the general mass behind it moves forward. Though the top of the flow quickly cools down, the molten underbelly of the flow is buffered by the solidifying rock above it, and by this mechanism lava flows can sustain movement for long periods of time. Pāhoehoe flows, in contrast, move in more conventional sheets, or by the advancement of lava "toes" in snaking lava columns. Increasing viscosity on the part of the lava or shear stress on the part of local topography can morph a pāhoehoe flow into an a'a one, but the reverse never occurs.

Although most shield volcanoes are by volume almost entirely Hawaiian and basaltic in origin, they are rarely exclusively so. Some volcanoes, like Mount Wrangell in Alaska and Cofre de Perote in Mexico, exhibit large enough swings in their historical magmatic eruptive characteristics to cast strict categorical assignment in doubt; one geological study of de Perote went so far as to suggest the term "compound shield-like volcano" instead. Most mature shield volcanoes have multiple cinder cones on their flanks, the results of tephra ejections common during incessant activity and markers of currently and formerly active sites on the volcano. One prominent such parasitic cones is Pu□u □Ō□ō on Kīlauea—continuous activity ongoing since 1983 has built up a 2,290 ft (698 m) tall cone at the site of one of the longest-lasting rift eruptions in known history.

The Hawaiian shield volcanoes and the Galápagos islands are unique in that they are not located near any plate boundaries; instead, the two chains are fed by the movement of oceanic plates over an upwelling of magma known as a hotspot. Over millions of years, the tectonic movement that moves continents also creates long volcanic trails across the seafloor. The Hawaiian and Galápagos shields, and other hotspot shields like them, are both constructed of oceanic island basalt. Their lavas are characterized by high levels of sodium, potassium, and aluminium.

Features common in shield volcanism include lava tubes. Lava tubes are cave-like volcanic straights formed by the hardening of overlaying lava. These structures help further the propagation of lava, as the walls of the tube insulates the lava within. Lava tubes can account for a large portion of shield volcano activity; for example, an estimated 58% of the lava forming Kilauea comes from lava tubes.

In some shield volcano eruptions, basaltic lava pours out of a long fissure instead of a central vent, and shrouds the countryside with a long band of volcanic material in the form of a broad plateau. Plateaus of this type exist in Iceland, Washington, Oregon, and Idaho; the most prominent ones are situated along the Snake River in Idaho and the Columbia River in Washington and Oregon, where they have been measured to be over a 1 mi (2 km) in thickness.

Calderas are a common feature on shield volcanoes. They are formed and reformed over the volcano's lifespan. Long eruptive periods form cinder cones, which then collapse over time to form calderas. The calderas are often filled up by future eruptions, or formed elsewhere, and this cycle of collapse and regeneration takes place throughout the volcano's lifespan.

Interactions between water and lava at shield volcanoes can cause some eruptions to become hydrovolcanic. These explosive eruptions are drastically different from the usual shield volcanic activity, and are especially prevalent at the waterbound volcanoes of the Hawaiian Isles.

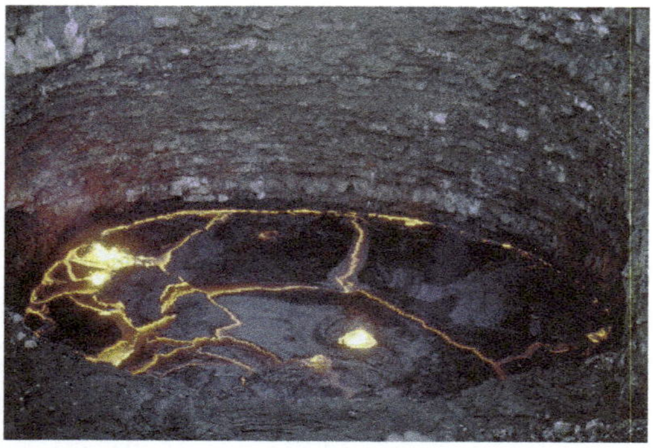

A lava lake in the caldera of Erta Ale, an active shield volcano in Ethiopia

The Thurston lava tube on Hawaii island, now a tourist attraction in the Hawaii Volcanoes National Park

Distribution

Shield volcanoes are found worldwide. They can form over hotspots (points where magma from below the surface wells up), such as the Hawaiian-Emperor seamount chain and the Galápagos Islands, or over more conventional rift zones, such as the Icelandic shields and the shield volcanoes of East Africa. Many shield volcanoes are found in ocean basins, such as Tamu Massif, the world's largest, although they can be found inland as well—East Africa being one example of this.

Hawaiian Islands

The largest and most prominent shield volcano chain in the world is the Hawaiian Islands, a chain of hotspot volcanoes in the Pacific Ocean. The Hawaiian volcanoes are characterized by frequent rift eruptions, their large size (thousands of km^3 in volume), and their rough, decentralized shape. Rift zones are a prominent feature on these volcanoes, and account for their seemingly random volcanic structure. They are fueled by the movement of the Pacific Plate over the Hawaii hotspot, and form a long chain of volcanoes, atolls, and seamounts 2,600 km (1,616 mi) long with a total volume of over 750,000 km^3 (179,935 cu mi). The chain contains at least 43 major volcanoes, and Meiji Seamount at its terminus near the Kuril-Kamchatka Trench is 85 million years old. The volcanoes follow a distinct evolutionary pattern of growth and death.

The chain includes the second largest volcano on Earth, Mauna Loa, which stands 4,170 m (13,680 ft) above sea level and reaches a further 13 km (8 mi) below the waterline and into the crust, approximately 80,000 km^3 (19,000 cu mi) of rock. Kilauea, meanwhile, is one of the most active volcanoes on Earth, with the current ongoing eruption having begun in January 1983.

Galápagos Islands

The Galápagos Islands are an isolated set of volcanoes, consisting of shield volcanoes and lava plateaus, located 1,200 km (746 mi) west of Ecuador. They are driven by the Galápagos hotspot, and are between approximately 4.2 million and 700,000 years of age. The largest island, Isabela Island, consists of six coalesced shield volcanoes, each delineated by a large summit caldera. Española, the oldest island, and Fernandina, the youngest, are also shield volcanoes, as are most of the other islands in the chain. The Galápagos Islands are perched on a large lava plateau known as the Galápagos Platform. This platform creates a shallow water depth of 360 to 900 m (1,181 to 2,953 ft) at the base of the islands, which stretch over a 174 mi (280 km)-long diameter. Since Charles Darwin's famous visit to the islands in 1835, there have been over 60 recorded eruptions in the islands, from six different shield volcanoes. Of the 21 emergent volcanoes, 13 are considered active.

Blue Hill is a shield volcano on the south western part of Isabela Island in the Galápagos Islands and is one of the most active in the Galapagos, with the last eruption between May and June 2008. The Geophysics Institute at the National Polytechnic School in Quito houses an international team of seismologists and volcanologists whose responsibility is to monitor Ecuadors numerous active volcanoes in the Andean Volcanic Belt and the Galapagos Islands. La Cumbre is an active shield volcano on Fernandina Island in the Galapagos that has been erupting since April 11, 2009.

The Galápagos islands are geologically young for such a big chain, and the pattern of their rift zones follows one of two trends, one north-northwest, and one east-west. The composition of the lavas of the Galápagos shields are strikingly similar to those of the Hawaiian volcanoes. Curiously, they do not form the same volcanic "line" associated with most hotspots. They are not alone in this regard; the Cobb-Eickelberg Seamount chain in the North Pacific is another example of such a delineated chain. In addition, there is no clear pattern of age between the volcanoes, suggesting a complicated, irregular pattern of creation. How exactly the islands were formed remains a geological mystery, although several theories have been fronted.

Iceland

Another major center of shield volcanic activity is Iceland. Located over the Mid-Atlantic Ridge, a divergent tectonic plate in the middle of the Atlantic Ocean, Iceland is the site of about 130 volcanoes of various types. Icelandic shield volcanoes are generally of Holocene age, between 5000 and 10000 years old, except for the island of Surtsey, a

Surtseyan shield. The volcanoes are also very narrow in distribution, occurring in two bands in the West and North Volcanic Zones. Like Hawaiian volcanoes, their formation initially begins with several eruptive centers before centralizing and concentrating at a single point. The main shield then forms, burying the smaller ones formed by the early eruptions with its lava.

Skjaldbreiður, Iceland, is eponymous for shield volcanoes.

Icelandic shields are mostly small (~15 km^3 (4 cu mi)), symmetrical (although this can affected by surface topography), and characterized by eruptions from summit calderas. They are composed of either tholeiitic olivine or picritic basalt. The tholeiitic shields tend to be wider and shallower than the picritic shields. They do not follow the pattern of caldera growth and destruction that other shield volcanoes do; caldera may form, but they generally do not disappear.

East Africa

East Africa is the site of volcanic activity generated by the development of the East African Rift, a developing plate boundary in Africa, and from nearby hotspots. Some volcanoes interact with both. Shield volcanoes are found near the rift and off the coast of Africa, although stratovolcanoes are more common. Although sparsely studied, the fact that all of its volcanoes are of Holocene age reflects how young the volcanic center is. One interesting characteristic of East African volcanism is a penchant for the formation of lava lakes; these semi-permanent lava bodies, extremely rare elsewhere, form in about nine percent of African eruptions.

The most active shield volcano in Africa is Nyamuragira. Eruptions at the shield volcano are generally centered within the large summit caldera or on the numerous fissures and cinder cones on the volcano's flanks. Lava flows from the most recent century extend down the flanks more than 30 km (19 mi) from the summit, reaching as far as Lake Kivu. Erta Ale in Ethiopia is another active shield volcano, and one of the few places in the world with a permanent lava lake, which has been active since at least 1967, and possibly since 1906. Other volcanic centers include Menengai, a massive shield caldera, and Mount Marsabit, near the town of Marsabit.

Extraterrestrial Volcanoes

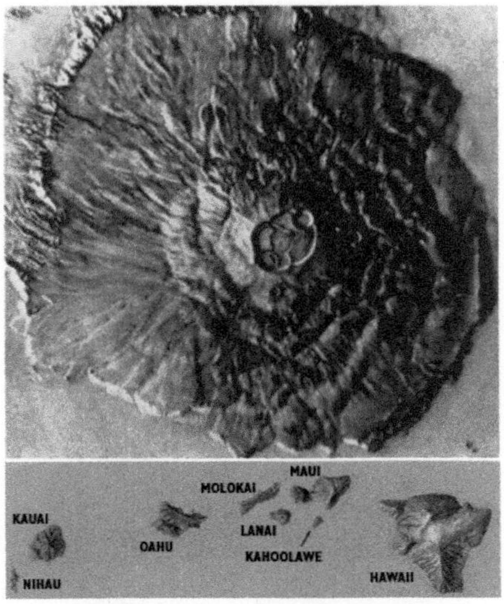

Scaled image showing Olympus Mons, top, and the Hawaiian island chain, bottom. Martian volcanoes are far larger than those found on Earth.

Volcanoes are not limited to Earth; they can exist on any rocky planet or moon large or active enough to have a molten inner core, and since probes were first launched in the 1960s, volcanoes have been found across the solar system. Shield volcanoes and volcanic vents have been found on Mars, Venus, and Io; cryovolcanoes on Triton; and subsurface hotspots on Europa.

The volcanoes of Mars are very similar to the shield volcanoes on Earth. Both have gently sloping flanks, collapse craters along their central structure, and are built of highly fluid lavas. Volcanic features on Mars were observed long before they were first studied in detail during the 1976–1979 Viking mission. The principal difference between the volcanoes of Mars and those on Earth is in terms of size; Martian volcanoes range in size up to 14 mi (23 km) high and 370 mi (595 km) in diameter, far larger than the 6 mi (10 km) high, 74 mi (119 km) wide Hawaiian shields. The highest of these, Olympus Mons, is the tallest known mountain on any planet in the solar system.

Venus also has over 150 shield volcanoes which are much flatter, with a larger surface area than those found on Earth, some having a diameter of more than 700 km (430 mi). Although the majority of these are long extinct it has been suggested, from observations by the Venus Express spacecraft, that many may still be active.

Mud Volcano

Mud volcano or mud dome refers to formations created by geo-exuded mud or slurries, water and gases. There are several geological processes that may cause the formation of mud volcanoes. Mud volcanoes are not true igneous volcanoes as they produce no lava. The Earth continuously exudes a mud-like substance, which may sometimes be referred to as a "mud volcano". Mud volcanoes may range in size from merely 1 or 2 meters high and 1 or 2 meters wide, to 700 meters high and 10 kilometers wide. Smaller mud exudations are sometimes referred to as mud-pots. The largest (man made) mud volcano is "Lusi" in Java, Indonesia, which is 10 kilometres (6 mi) in diameter.

A series of mud volcanoes in Gobustan, Azerbaijan

El Totumo mud volcano in Santa Catalina, Colombia.

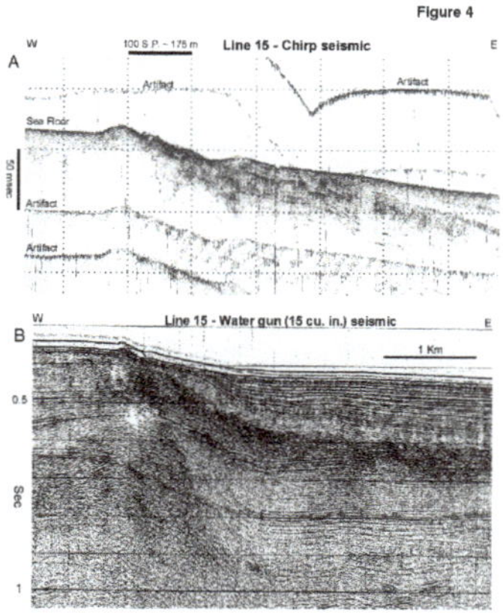

Mud volcano in the Gulf of Mexico sea bottom

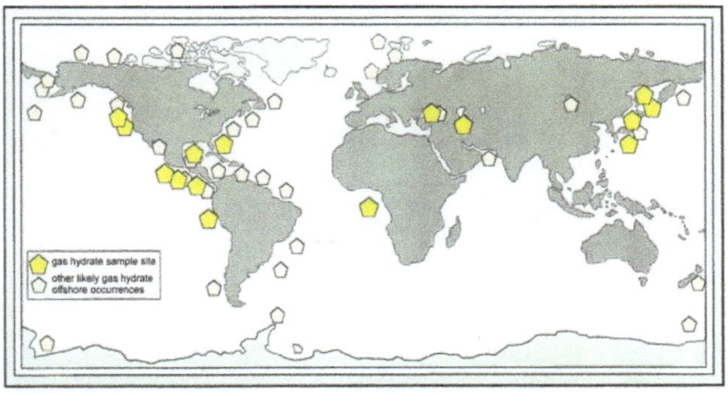

Hydrate-bearing sediments, which often are associated with mud volcano activity. Source: USGS, 1996.

The mud produced by mud volcanoes is most typically formed as hot water, which has been heated deep below the earth's surface, begins to mix and blend with subterranean mineral deposits, thus creating the mud slurry exudate. This material is then forced upwards through a geological fault or fissure due to local subterranean pressure imbalances. Mud volcanoes are associated with subduction zones and about 1100 have been identified on or near land. The temperature of any given active mud volcano generally remains fairly steady and is much lower than the typical temperatures found in igneous volcanoes. Mud volcano temperatures can range from near 100 °C (212 °F) to occasionally 2 °C (36 °F), some being used as popular "mud baths."

About 86% of the gas released from these structures is methane, with much less carbon dioxide and nitrogen emitted. Ejected materials are most often a slurry of fine solids suspended in water that may contain a mixture of salts, acids and various hydrocarbons.

Possible mud volcanoes have been identified on Mars.

Details

A mud volcano may be the result of a piercement structure created by a pressurized mud diapir that breaches the Earth's surface or ocean bottom. Their temperatures may be as low as the freezing point of the ejected materials, particularly when venting is associated with the creation of hydrocarbon clathrate hydrate deposits. Mud volcanoes are often associated with petroleum deposits and tectonic subduction zones and orogenic belts; hydrocarbon gases are often erupted. They are also often associated with lava volcanoes; in the case of such close proximity, mud volcanoes emit incombustible gases including helium, whereas lone mud volcanoes are more likely to emit methane.

Approximately 1,100 mud volcanoes have been identified on land and in shallow water. It has been estimated that well over 10,000 may exist on continental slopes and abyssal plains.

Features

- Gryphon: steep-sided cone shorter than 3 meters that extrudes mud

- Mud cone: high cone shorter than 10 meters that extrudes mud and rock fragments
- Scoria cone: cone formed by heating of mud deposits during fires
- Salse: water-dominated pools with gas seeps
- Spring: water-dominated outlets smaller than 0.5 metres
- Mud shield

Emissions

Most liquid and solid material is released during eruptions, but seeps occur during dormant periods.

The mud is rich in halite (rock salt).

First-order estimates of mud volcano emissions have been made (1 Tg = 1 million metric tonnes).

- 2002: L. I. Dimitrov estimated that 10.2–12.6 Tg/yr of methane is released from onshore and shallow offshore mud volcanoes.
- 2002: Etiope and Klusman estimated at least 1–2 and as much as 10–20 Tg/yr of methane may be emitted from onshore mud volcanoes.
- 2003: Etiope, in an estimate based on 120 mud volcanoes: "The emission results to be conservatively between 5 and 9 Tg/yr, that is 3–6% of the natural methane sources officially considered in the atmospheric methane budget. The total geologic source, including MVs (this work), seepage from seafloor (Kvenvolden et al., 2001), microseepage in hydrocarbon-prone areas and geothermal sources (Etiope and Klusman, 2002), would amount to 35–45 Tg/yr."
- 2003: analysis by Milkov et al. suggests that the global gas flux may be as high as 33 Tg/yr (15.9 Tg/yr during quiescent periods plus 17.1 Tg/yr during eruptions). Six teragrams per year of greenhouse gases are from onshore and shallow offshore mud volcanoes. Deep-water sources may emit 27 Tg/yr. Total may be 9% of fossil CH_4 missing in the modern atmospheric CH_4 budget, and 12% in the preindustrial budget.
- 2003: Alexei Milkov estimated approximately 30.5 Tg/yr of gases (mainly methane and CO_2) may escape from mud volcanoes to the atmosphere and the ocean.
- 2003: Achim J. Kopf estimated 1.97×10^{11} to 1.23×10^{14} m^3 of methane is released by all mud volcanoes per year, of which 4.66×10^{7} to 3.28×10^{11} m^3 is from surface volcanoes. That converts to 141–88,000 Tg/yr from all mud volcanoes, of which 0.033–235 Tg is from surface volcanoes.

Locations

Europe

Dozens of mud volcanoes can be found on the Taman Peninsula of Russia and the Kerch Peninsula of the Crimean disputed territories. In Italy, they are common in the northern front of the Apennines (e.g., Salse di Nirano) and in Sicily. On August 24, 2013, a mud volcano appeared in the centre of the via Coccia di Morto roundabout in Fiumicino near Rome.

Two mud volcanoes on the Taman Peninsula near Taman Stanitsa

Another relatively accessible place where mud volcanoes can be found in Europe are the Berca Mud Volcanoes near Berca in Buzău County, Romania, close to the Carpathian Mountains.

Asia

Indonesia

Drilling may have resulted in the Sidoarjo mud flow on May 29, 2006, or an earthquake in the Porong subdistrict of East Java province, Indonesia. The mud covered about 440 hectares, 1,087 acres (4.40 km^2) (2.73 mi^2), and inundated four villages, homes, roads, rice fields, and factories, displacing about 24,000 people and killing 14. The gas exploration company involved was operated by PT Lapindo Brantas and the earthquake that may have triggered the mud volcano was the 6.3 magnitude Yogyakarta earthquake of May 27, 2006. In 2008, it was termed the world's largest mud volcano. It is beginning to show signs of catastrophic collapse, according to geologists who have been monitoring it and the surrounding area. A catastrophic collapse could sag the vent and surrounding area by up to 150 metres (490 ft) in the next decade. In March 2008, the scientists observed drops of up to 3 metres (9.8 ft) in one night. Most of the subsidence in the area around the volcano is more gradual, at around 1 millimetre (0.039 in) per day. A study by a group of Indonesian geo-scientists led by Bambang Istadi predicted the area affected by the mudflow over a ten-year period. More recent studies carried out in 2011 predict that the mud will flow for another 20 years, or even longer. Now named Lusi – a contraction of *Lumpur Sidoarjo*, where *lumpur* is the Indonesian word for "mud" – the mud volcano appears to be a hydrocarbon/hydrothermal hybrid.

Central Asia

Many mud volcanoes exist on the shores of the Black Sea and Caspian Sea. Tectonic forces and large sedimentary deposits around the latter have created several fields of mud volcanoes, many of them emitting methane and other hydrocarbons. Features over 200 metres (656 ft) high are in Azerbaijan, with large eruptions sometimes producing flames of similar scale.

Satellite image of mud volcanoes in Pakistan

Iran and Pakistan possess mud volcanoes in the Makran range of mountains in the south of the two countries. The world's largest and highest mud volcano is located in Balochistan, Pakistan. It is known as *Baba Chandrakup* (literally Father Moonwell) on the way to Hinglaj. It is a pilgrim site for Hindus.

There are mud volcanoes in Georgia, such as at Akhtala.

Azerbaijan

Mud volcanoes in Gobustan, Azerbaijan

Azerbaijan and its Caspian coastline are home to nearly 400 mud volcanoes, more than half the total throughout the continents. In 2001, one mud volcano 15 kilometres (9 mi) from Baku made world headlines when it started ejecting flames 15 metres (49 ft) high.

In Azerbaijan, eruptions are driven from a deep mud reservoir which is connected to the surface even during dormant periods, when seeping water shows a deep origin. Seeps have temperatures that are generally above ambient ground temperature by 2 °C (3.6 °F) - 3 °C (5.4 °F). Lökbatan Mud Volcano since 1998 submitted to UNESCO's World Heritage Site tentative list.

Iran

There are many mud volcanoes in Iran: in Hormozgan province, Sistan and Baluchestan Province and Golestan Province.

Mud volcano in Hormozgan, south of Iran

India

The island of Baratang, part of the Great Andaman archipelago in the Andaman Islands, Indian Ocean, has several sites of mud volcanic activity. There was a significant eruption event in 2003.

Pakistan

In Pakistan there are more than 80 active mud volcanoes in Balochistan province; there are about 10 locations with clusters of mud volcanoes. In the west, in Gwadar District, the mud volcanoes are very small and mostly sit in the south of Jabal-e-Mehdi toward Sur Bandar. Many more are in the northeast of Ormara. The remainder are in Lasbela District and are scattered between south of Gorangatti on Koh Hinglaj to Koh Kuk in the North of Miani Hor in the Hangol Valley. In this region, the heights of mud volcanoes range between 800 to 1,550 feet (243.8 to 472.4 m). The most famous is Chandragup. The biggest crater found at 25°33'13.63"N. 65°44'09.66"E is about 450 feet (137.16 m) in diameter. Most mud volcanoes in this region are in out-of-reach areas having very difficult terrain. Dormant mud volcanoes stand like columns of mud in many other areas.

Philippines

In the Turtle Islands, in the province of Tawi-Tawi, the southwestern edge of the Philippines bordering Malaysia, presence of mud volcanoes are evident on three of the islands – Lihiman, Great Bakkungan and Boan Islands. The northeastern part of Lihiman Island is distinguished for having more violent kind of mud extrusions mixed with large pieces of rocks, creating a 20-m (66-ft) wide crater on that hilly part of the island. Such extrusions are reported to be accompanied by mild earthquakes and evidence of extruded materials can be found high in the surrounding trees. Submarine mud extrusions off the island have been observed by local residents.

Other Asian Locations

Mud volcano landscape at Oesilo, Oecusse District, Timor-Leste

- China has a number of mud volcanoes in Xinjiang province.
- Some active mud volcanoes are in Oesilo (Oecusse District, East Timor). Arthur Wichmann reports a mud volcano in Bibiluto (Viqueque District), which erupted between 1856 and 1879.
- There are mud volcanoes at the Minn Buu Township, Magway division in Myanmar (Burma).
- There are two active mud volcanoes in South Taiwan and several inactive ones. The Wushan Mud Volcanoes (烏山頂泥火山 in Chinese) are in the Yanchao District of Kaohsiung City. There are active mud volcanoes in Wandan township of Pingtung county.
- There are mud volcanoes on the island of Pulau Tiga, off the western coast of the Malaysian state of Sabah on Borneo.
- A drilling accident offshore of Brunei on Borneo in 1979 caused a mud volcano which took 20 relief wells and nearly 30 years to halt.

North America

A cold mud pot in Northern California, showing the scale

A cold mud pot in Glenblair, California

Aerial view of mud volcanoes on the Nahlin Plateau, British Columbia. Scale - each volcano approximately 20 m in diameter.

Mud volcanoes of the North American continent include:

- A field of small (<2 metres (6.6 ft) high) fault-controlled, cold mud volcanoes is on California's Mendocino Coast, near Glenblair and Fort Bragg, California. The fine grained clay is occasionally harvested by local potters.
- Shrub and Klawasi mud volcanoes in the Copper River basin by the Wrangell Mountains, Alaska. Emissions are mostly CO_2 and nitrogen; the volcanoes are associated with magmatic processes.

- An unnamed mud volcano 30 metres (98 ft) high and with a top about 100 metres (328 ft) wide, 24 kilometres (15 mi) off Redondo Beach, California, and 800 metres (2,620 ft) under the surface of the Pacific Ocean.
- A field of small (<3 metres (9.8 ft)) mud volcanoes in the Salton Sea geothermal area near the town of Niland, California. Emissions are mostly CO_2.
- Smooth Ridge mud volcano in 1,000 metres (3,280 ft) of water near Monterey Canyon, California.
- Kaglulik mud volcano, 43 metres (141 ft) under the surface of the Beaufort Sea, near the northern boundary of Alaska and Canada. Petroleum deposits are believed to exist in the area.
- Maquinna mud volcano, located 16–18 kilometres (9.9–11.2 mi) west of Vancouver Island, British Columbia, Canada.
- There are many mud volcanoes in Trinidad and Tobago in the Caribbean, near oil reserves in southern parts of the island of Trinidad. As of August 15, 2007, the mud volcano titled the Moruga Bouffle was said to being spitting up methane gas which shows that it is active. There are several other mud volcanoes in the tropical island which include:
    - the Devils Woodyard mud volcano near Trinidad and Tobago
    - the Moruga Bouffe mud volcano near Moruga
    - the Piparo mud volcano
    - the Chatham mud volcano underwater in the Columbus Channel; this mud volcano periodically produces a short-lived island.

Yellowstone's "Mud Volcano"

The name of Yellowstone National Park's "Mud Volcano" feature and the surrounding area is misleading; it consists of hot springs, mud pots and fumaroles, rather than a true mud volcano. Depending upon the precise definition of the term *mud volcano*, the Yellowstone formation could be considered a hydrothermal mud volcano cluster. The feature is much less active than in its first recorded description, although the area is quite dynamic. Yellowstone is an active geothermal area with a magma chamber near the surface, and active gases are chiefly steam, carbon dioxide, and hydrogen sulfide.

However, there are mud volcanoes and mud geysers elsewhere in Yellowstone. One, the "Vertically Gifted Cyclic Mud Pot" sometimes acts as a geyser, throwing mud up to 30 feet high.

The mud volcano in Yellowstone was previously a mound until suddenly it tore itself apart into the formation seen today.

South America

Venezuela

Yagrumito mud volcano in Monagas, Venezuela (6 km from Maturín)

The eastern part of Venezuela contains several mud volcanoes(a.k.a. Mud dome), all of them having an origin related to oil deposits. The mud of 6 kilometres (3.7 mi) from Maturín, Venezuela contains water, biogenic gas, hydrocarbons and an important quantity of salt. Cows from the savanna often gather around to lick the dried mud for its salt content, which is an integral part of their diet needed to produce milk.

Trinidad

Photo of mud volcano in southern Trinidad c. 1967

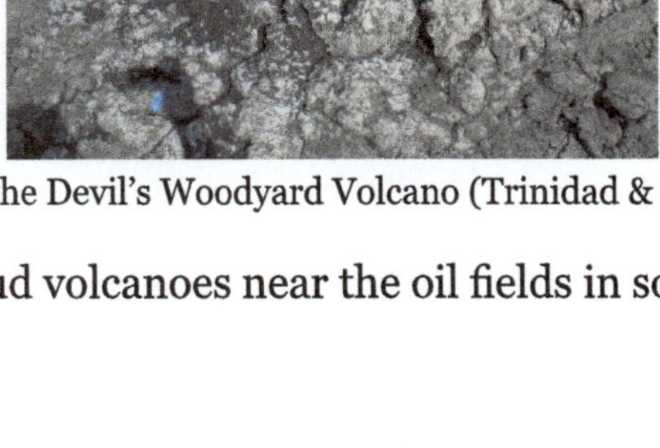

One of the Devil's Woodyard Volcano (Trinidad & Tobago)

There are a number of mud volcanoes near the oil fields in southern Trinidad.

Colombia

Volcan El Totumo, which marks the division between Bolívar and Atlantico in Colombia. This volcano is approximately 50 feet (15 m) high and can accommodate 10 to 15 people in its crater; many tourists and locals visit this volcano due to the alleged medicinal benefits of the mud; it is next to a *cienaga*, or lake. This volcano is under legal dispute between the Bolivar and Atlantico Departamentos because of its tourist value.

Supervolcano

Map of known supervolcanoes around the world:
Volcanic Explosivity Index (VEI) 8
Volcanic Explosivity Index (VEI) 7

A supervolcano is any volcano capable of producing a volcanic eruption with an ejecta mass greater than 10^{15} kg (10^{12} t). Supervolcanoes occur when magma in the mantle rises into the crust but is unable to break through the crust, and pressure builds in a large and growing magma pool until the crust is unable to contain the pressure. This can occur at hotspots (for example, Yellowstone Caldera) or at subduction zones (for example, Toba).

Another setting for the eruption of very large amounts of volcanic material is in large igneous provinces, which can cover huge areas with lava and volcanic ash, causing long-lasting climate change (such as the triggering of a small ice age or global warming), which can threaten species with extinction.

Terminology

The origin of the term *supervolcano* is linked to an early 20th-century scientific debate about the geological history and features of the Three Sisters volcanic region of Oregon in the United States. In 1925, Edwin T. Hodge suggested that a very large volcano, which he named Mount Multnomah, had existed in that region. He believed that several peaks in the Three Sisters area are the remnants left after Mount Multnomah had been largely destroyed by violent volcanic explosions, similar to Mount Mazama. In 1948, the possible existence of Mount Multnomah was ignored by volcanologist Howel Williams in his book *The Ancient Volcanoes of Oregon*. The book was reviewed in 1949 by another volcanologist, F. M. Byers Jr. In the review, Byers refers to Mount Multnomah as a supervolcano. Although Hodge's suggestion that Mount Multnomah is a supervolcano was rejected long ago, the term *supervolcano* was popularised by the BBC popular science television program *Horizon* in 2000 to refer to eruptions that produce extremely large amounts of ejecta.

Volcanologists and geologists do not refer to "supervolcanoes" in their scientific work, since this is a blanket term that can be applied to a number of different geological settings. Since about 2000, however, the term has been used by professionals for communicating science to the public. The term *megacaldera* is sometimes used for caldera supervolcanoes, such as the Blake River Megacaldera Complex in the Abitibi greenstone belt of Ontario and Quebec, Canada. Eruptions that rate VEI 8 are termed *super eruptions*. Though there is no well-defined minimum explosive size for a "supervolcano", there are at least two types of volcanic eruptions that have been identified as supervolcanoes: large igneous provinces and massive eruptions.

Large Igneous Provinces

Large igneous provinces (LIP) such as Iceland, the Siberian Traps, Deccan Traps, and the Ontong Java Plateau are extensive regions of basalts on a continental scale resulting from flood basalt eruptions. When created, these regions often occupy several thousand square kilometres and have volumes on the order of millions of cubic kilometers. In most cases, the lavas are normally laid down over several million years. They release large amounts of gases. The Réunion hotspot produced the Deccan Traps about 66 million years ago, coincident with the Cretaceous–Paleogene extinction event. The scientific consensus is that a meteor impact was the cause of the extinction event, but the volcanic activity may have caused environmental stresses on extant species up to the Cretaceous–Paleogene boundary. Additionally, the largest flood basalt event (the Siberian Traps) occurred around 250 million years ago and was coincident with the largest mass extinction in history, the Permian–Triassic extinction event, although it is also unknown whether it was completely responsible for the extinction event.

Such outpourings are not explosive, though lava fountains may occur. Many volcanologists consider that Iceland may be a LIP that is currently being formed. The last major outpouring occurred in 1783–84 from the Laki fissure which is approximately 40 km (25 mi) long. An estimated 14 km^3 (3.4 cu mi) of basaltic lava was poured out during the eruption.

The Ontong Java Plateau now has an area of about 2,000,000 km^2 (770,000 sq mi), and the province was at least 50% larger before the Manihiki and Hikurangi Plateaus broke away.

Massive Explosive Eruptions

Volcanic eruptions are classified using the Volcanic Explosivity Index, or VEI. It is a logarithmic scale, which means that an increase of one in VEI number is equivalent to a tenfold increase in volume of erupted material. VEI 7 or VEI 8 eruptions are so powerful that they often form circular calderas rather than cones because the downward withdrawal of magma causes the overlying rock mass to collapse into the empty magma chamber beneath it.

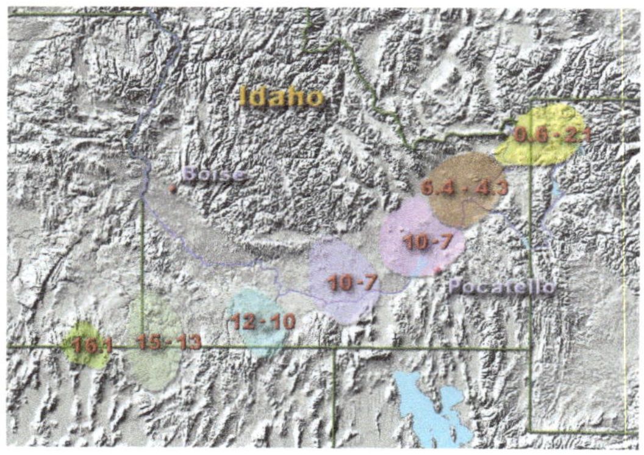

Location of Yellowstone hotspot over time (numbers indicate millions of years before the present).

VEI 8 eruptions are colossal events that throw out at least 1,000 km³ (240 cu mi) bulk volume.

Satellite image of Lake Toba, the site of a VEI 8 eruption ~75,000 years ago.

VEI 7 events eject bulk volume at least 100 km³ (24 cu mi).

VEI 6 eruptions occurred at Krakatoa in 1883 and Mount Pinatubo in 1991. These eruptions ejected ~10 and 25 km³ (2.4 and 6.0 cu mi) bulk volume, respectively. At Krakatoa, the Dutch colonial authorities claimed that the death toll was 36,417, but other estimates consider that the death toll is in excess of 120,000.

The 1980 Mount St. Helens eruption was a VEI 5 eruption, with 1.2 km³ (0.29 cu mi) bulk volume of ejecta.

Known Supereruptions

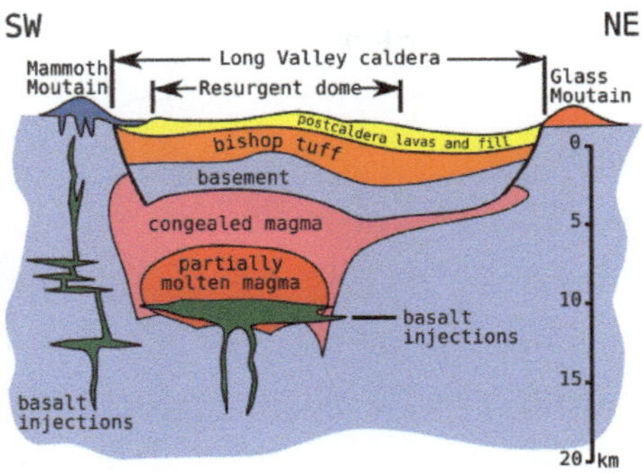

Cross-section through Long Valley Caldera.

VEI 8

This list is incomplete; you can help by expanding it.

Based on incomplete statistics, at least 60 VEI 8 eruptions have been identified.

VEI 7

VEI 7 eruptions, less colossal but still supermassive, have occurred in historical times. The only ones in the past 2,000 years are Tambora, in 1815, Taupo Volcano's Hatepe eruption, c. 232, Baekdu Mountain, in 946–47, and the eruption of Mount Samalas in 1257.

Media Portrayal

- In 2004, Naked Science TV show aired supervolcano on National Geographic Channel.
- In 2005, a two-part television disaster film called *Supervolcano* aired on BBC One, the Discovery Channel, and other television networks worldwide.
- *Nova* featured an episode *Mystery of the Megavolcano* in September 2006 examining such eruptions in the last 100,000 years.
- In 2006, the Sci Fi Channel aired the documentary *Countdown to Doomsday* which featured a segment called "Supervolcano". The same year, ABC News aired the documentary *Last Days on Earth*, which featured a segment called "Supervolcano".
- Also in 2006, the Syfy Channel series Stargate Atlantis episode entitled "Inferno" featured a supervolcano as the major plot device. Dr. Rodney McKay, one of the main characters, uses Yellowstone National Park to describe what a supervolcano is.
- In the episode "Humanity" of the television drama Young Justice, the team must relieve the pressure of the Yellowstone Caldera supervolcano caused by Red Vol-

cano before an eruption with the potential for mass extinction takes place.

- In 2009, the apocalypse-themed film *2012* featured the super-eruption of the massive Yellowstone Caldera, a result of the Earth's core heating up. This made most of the United States uninhabitable.
- In 2010, the SyFy series Warehouse 13 featured an episode entitled *Reset* in which a supervolcano, specifically the Yellowstone Caldera, plays an important role.
- In December 2011, author Harry Turtledove published *Supervolcano: Eruption*, the first of a planned four-novel series about events leading up to and following a fictional eruption of the Yellowstone Caldera. The second book in the series, *Supervolcano: All Fall Down*, was published in December 2012. The third book *Supervolcano: Things Fall Apart*, was published in December 2013.
- At the end of Terry Pratchett and Stephen Baxter's novel *Long War* the Yellowstone Caldera erupts. The aftermath is featured in the next volume, *Long Mars*.
- The Siberian Traps and Lake Toba were featured in Animal Planet's prehistoric documentary *Animal Armageddon* (each in one episode), where speculation of the lifeforms they afflicted were brought to life.

Satellite image of San Salvador, El Salvador and the Lake Ilopango caldera (also known as the Dark Age volcano) in the Valley of the Hammocks, site of a VEI 6-8 eruption and said to be ground zero for the extreme weather events of 535–536, when a dark veil settled on the world.

Volcano, lake, and caldera locations in the Taupo Volcanic Zone

Submarine Volcano

Submarine volcanoes are underwater vents or fissures in the Earth's surface from which magma can erupt. A large number of submarine volcanoes are located near areas of tectonic plate movement, known as mid-ocean ridges. The volcanoes at mid-ocean ridges alone are estimated to account for 75% of the magma output on Earth. Although most submarine volcanoes are located in the depths of seas and oceans, some also exist in shallow water, and these can discharge material into the atmosphere during an eruption. The total number of submarine volcanoes is estimated to be over 1 million, of which some 75 000 rise more than 1 km above the seabed.

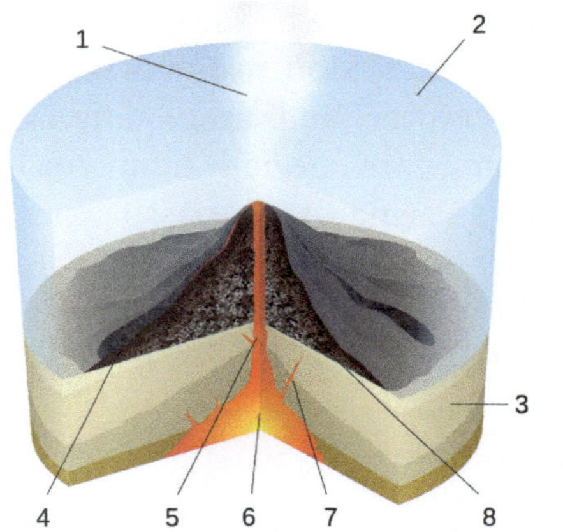

Scheme of a submarine eruption.
1. Water vapor cloud
2. Water
3. Stratum
4. Lava flow
5. Magma conduit
6. Magma chamber
7. Dike
8. Pillow lava

Hydrothermal vents, sites of abundant biological activity, are commonly found near submarine volcanoes.

Effect of Water on Volcanoes

The presence of water can greatly alter the characteristics of a volcanic eruption and the explosions of underwater volcanoes in comparison to those on land.

For instance, water causes magma to cool and solidify much more quickly than in a terrestrial eruption, often turning it into a volcanic glass. The lava formed by submarine volcanoes is quite different from terrestrial lava. Upon contact with water, a solid crust forms around the lava. Advancing lava flows into this crust, forming what is known as pillow lava.

Below ocean depths of about 2200 m, where the pressure exceeds the critical pressure of water (22.06 MPa or about 218 atmospheres), it does not longer boil; it becomes a supercritical fluid. Without boiling sounds, deep-sea volcanoes can be difficult to detect at great distances using hydrophones.

Topographical Activity

Scientists still have much to learn about the location and activity of underwater volcanoes. The Kolumbo underwater volcano in the Aegean Sea was discovered in 1650 when it burst from the sea and erupted, killing 70 people on the nearby island of Santorini. More recently, NOAA's Office of Ocean Exploration has funded exploration of submarine volcanoes, with the Ring of Fire missions to the Mariana Arc in the Pacific Ocean being particularly noteworthy. Using Remote Operated Vehicles, scientists studied underwater eruptions, ponds of molten sulfur, black smoker chimneys and even marine life adapted to this deep, hot environment.

Many submarine volcanoes are seamounts, typically extinct volcanoes that rise abruptly from a seafloor of 1,000 - 4,000 meters depth. They are defined by oceanographers as independent features that rise to at least 1,000 meters above the seafloor. The peaks are often found hundreds to thousands of meters below the surface, and are therefore considered to be within the deep sea. An estimated 30,000 seamounts occur across the globe, with only a few having been studied. However, some seamounts are also unusual. For example, while the summits of seamounts are normally hundreds of meters below sea level, the Bowie Seamount in Canada's Pacific waters rises from a depth of about 3,000 meters to within 24 meters of the sea surface.

Identifying types of Eruptions by Sounds

Deepest ever filmed submarine volcano, West Mata, May 2009.

There are two types of submarine eruptions: One is created by the slow release and bursting of large lava bubbles, and the other one is created by a quick explosion of gas bubbles. Lava can affect marine animals and ecosystems differently than gas can, so it is important to be able to distinguish the two.

Scientists have been able to connect sounds to sights. in both types of eruptions. In 2009, a video camera and a hydrophone were floating 1,200 meters below sea level in the Pacific Ocean near Samoa, watching and listening as the West Mata Volcano erupted in several ways. Putting video and audio together let researchers learn the sounds made by slow lava bursting and the different noises made by hundreds of gas bubbles.

References

- Barlow, Nadine (2008). Mars : an introduction to its interior, surface and atmosphere. Cambridge, UK: Cambridge University Press. ISBN 9780521852265.
- World Book: U · V · 20. Chicago, Illinois: Scott Fetzer. 2009. pp. 438–443. ISBN 978-0-7166-0109-8. Retrieved 22 August 2010.
- James S. Monroe; Reed Wicander (2006). The changing Earth : exploring geology and evolution (5th ed.). Belmont, CA: Brooks/Cole. p. 115. ISBN 978-0-495-55480-6. Retrieved February 22, 2011.
- Whittlesey, Lee (1995) [1995]. Death in Yellowstone: Accidents and Foolhardiness in the First National Park. Lanham, Maryland: Roberts Rinehart Publishers. ISBN 1-57098-021-7.
- Martin R. Speight, Peter A. Henderson, "Marine Ecology: Concepts and Applications", John Wiley & Sons, 2013. ISBN 978-1-4051-2699-1.
- Lambert, Maurice B. (1978). Volcanoes. North Vancouver, British Columbia: Energy, Mines and Resources Canada. ISBN 0-88894-227-3.
- John Watson (1 March 2011). "Principal Types of Volcanoes". United States Geological Survey. Retrieved 30 December 2013.
- Bruce Perry. "Igneous Rocks Tour: Volcanoes and Lava Flows". California State University, Long Beach. Retrieved 30 December 2013.
- Brian Clark Howard (5 September 2013). "New Giant Volcano Below Sea Is Largest in the World". National Geographic. Retrieved 31 December 2013.
- "How High is Mauna Loa?". Hawaiian Volcano Observatory—United States Geological Survey. 20 August 1998. Retrieved 5 February 2013.
- Navin Singh Khadka (28 February 2012). "Nepal in new bid to finally settle Mount Everest height". BBC News. Retrieved 10 December 2012.
- Byers, Jr., F. M. (1949) Reviews: The Ancient Volcanoes of Oregon by Howel Williams, The Journal of Geology, volume 57, number 3, May 1949, page 324. Retrieved 2012-08-17.
- Global Volcanism Program | Volcanoes of the World | Large Holocene Eruptions. Volcano.si.edu. Retrieved on 2011-11-18.
- "Volcanoes of South America: Galápagos Islands". Global Volcanism Program. Smithsonian National Museum of Natural History. Retrieved 22 February 2011.
- Ruth Andrews & Agust Gudmundsson (2006). "Holocene shield volcanoes in Iceland" (PDF). University of Göttingen. Retrieved 21 February 2011.
- Watson, John (February 5, 1997). "Extraterrestrial Volcanism". United States Geological Survey. Retrieved February 13, 2011.
- Dennis Normile (24 February 2011). "Indonesia's Infamous Mud Volcano Could Outlive All of Us". Science (AAAS). Retrieved 15 April 2011.
- "Evolution of Hawaiian Volcanoes". Hawaiian Volcano Observatory - United States Geological Survey. 8 September 1995. Retrieved 28 February 2011.
- "Summary of the Pu`u `Ō `ō-Kupaianaha Eruption, 1983-present". United States Geological Service - Hawaii Volcano Observatory. 4 October 2008. Retrieved 5 February 2011.

4

Major Features of Volcanoes

The major features of volcanoes covered in this chapter are fissure vent, lava dome and volcanic cone. Volcanic cones are built from the ejected particles from a volcanic vent, which subsequently piles around the vent in the shape of a cone, whereas fissure vents are linear volcanic vents through which lava erupts but usually without exploding.

Fissure Vent

A fissure vent, also known as a volcanic fissure or eruption fissure, is a linear volcanic vent through which lava erupts, usually without any explosive activity. The vent is often a few meters wide and may be many kilometers long. Fissure vents can cause large flood basalts which run first in lava channels and later in lava tubes. After some time the eruption builds up spatter resp. ash cones and may concentrate on one or some of them. Small fissure vents may not be easily discernible from the air, but the crater rows or the canyons built up by some of them are.

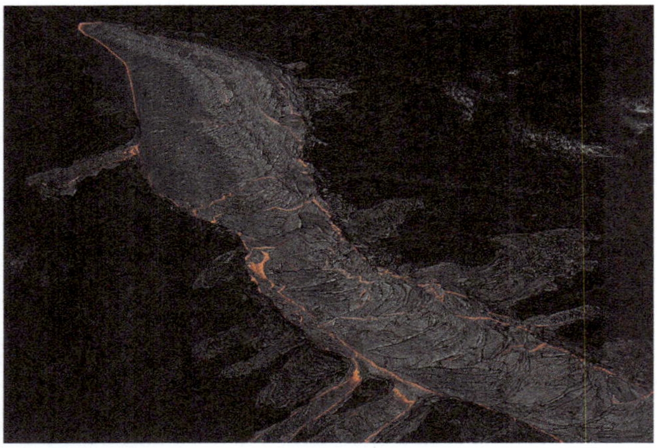

Lava channel on Hawaii

A fissure vent, also known as a volcanic fissure or eruption fissure, is a linear volcanic vent through which lava erupts, usually without any explosive activity. The vent is often a few meters wide and may be many kilometers long. Fissure vents can cause large flood basalts which run first in lava channels and later in lava tubes. After some time the eruption builds up spatter resp. ash cones and may concentrate on one or some of them. Small fissure vents may not be easily discernible from the air, but the crater rows or the canyons built up by some of them are.

Eldhraun, a lava field produced by the Laki craters

The dikes that feed fissures reach the surface from depths of a few kilometers and connect them to deeper magma reservoirs, often under volcanic centers. Fissures are usually found in or along rifts and rift zones, such as Iceland and the East African Rift. Fissure vents are often part of the structure of shield volcanoes.

Iceland

In Iceland, volcanic vents, which can be long fissures, often open parallel to the rift zones where the Eurasian and the North American Plate lithospheric plates are diverging, a system which is part of the Mid-Atlantic Ridge. Renewed eruptions generally occur from new parallel fractures offset by a few hundred to thousands of metres from the earlier fissures. This distribution of vents and sometimes voluminous eruptions of fluid basaltic lava usually build up a thick lava plateau rather than a single volcanic edifice. But there are also the central volcanoes, composite volcanoes, often with calderas, which have been formed during thousands of years and eruptions with one or more magma reservoirs undeneath controlling their resp. fissure system.

The Laki fissures, part of the Grímsvötn volcanic system, produced one of the biggest effusive eruptions on earth in historical times, in the form of a flood basalt of 12–14 km^3 of lava. During the Eldgjá eruption A.D. 934-40, another very big effusive fissure eruption in the volcanic system of Katla in South Iceland, ~18 km^3 (4.7 mi^3) of lava were released. In September 2014, a fissure eruption is ongoing on the site of the 18th century lava field Holuhraun. The eruption is part of an eruption series in the Bárðarbunga volcanic system.

Hawaii

The radial fissure vents of Hawaiian volcanoes also produce "curtains of fire" as lava fountains erupting along a portion of a fissure. These vents build up low ramparts of basaltic spatter on both sides of the fissure. More isolated lava fountains along the fissure produce crater rows of small spatter and cinder cones. The fragments that form a spatter cone are hot and plastic enough to weld together, while the fragments that form a cinder cone remain separate because of their lower temperature.

List of fissure vents

Name	Elevation		Location	Last eruption
	metres	feet	Coordinates	
Lakagígar	620	2034	64°04′N 18°14′W64.07°N 18.23°W	1783
Lanzarote	670	2198	29°02′N 13°38′W29.03°N 13.63°W	1824
Cordón Caulle	1798	5899	40°28′S 72°15′W40.46°S 72.25°W	2011
São Jorge Island	1053	3455	38°39′N 28°05′W38.65°N 28.08°W	1907
Quetena	5730	18799	22°15′S 67°25′W22.25°S 67.42°W	Unknown
Nejapa Miraflores	360	1181	12°07′N 86°19′W12.12°N 86.32°W	Unknown
Manda-Inakir	600+	1968	12°23′N 42°12′E12.38°N 42.20°E	1928
Hertali	900	2953	9°47′N 40°20′E9.78°N 40.33°E	Unknown
Estelí	899	2949	13°10′N 86°24′W13.17°N 86.40°W	Unknown
Butajiri Silti Field	2281	7484	8°03′N 83°51′E8.05°N 83.85°E	Unknown
Bishoftu Volcanic Field	1850+	6069	8°47′N 38°59′E8.78°N 38.98°E	Unknown
Holuhraun	730	2395	64°52′N 16°50′W64.87°N 16.83°W	2014
Alu	429	1407	13°49′N 40°33′E13.82°N 40.55°E	Unknown
Singu Plateau	507	1663	22°42′N 95°59′E22.70°N 95.98°E	Unknown
Ray Mountain	2050	6730	52°14′N 120°07′W52.23°N 120.12°W	Pleistocene
Eldgjá	800	2625	63°53′N 18°46′W63.88°N 18.77°W	934

Lava Dome

In volcanology, a lava dome or volcanic dome is a roughly circular mound-shaped pro-trusion resulting from the slow extrusion of viscous lava from a volcano. The geochem-istry of lava domes can vary from basalt to rhyolite although most preserved domes tend to have high silica content. The characteristic dome shape is attributed to high vis-cosity that prevents the lava from flowing very far. This high viscosity can be obtained in two ways: by high levels of silica in the magma, or by degassing of fluid magma. Since viscous basaltic and andesitic domes weather fast and easily break apart by further input of fluid lava, most of the preserved domes have high silica content and consist of rhyolite or dacite.

Image of the rhyolitic lava dome of Chaitén Volcano during its 2008–2010 eruption.

One of the Mono Craters, an example of a rhyolite dome.

Existence of lava domes has been suggested for some domed structures on the Martian surface in the western part of Arcadia Planitia and within Terra Sirenum.

Dome Dynamics

Lava domes in the crater of Mount St. Helens

Lava domes evolve unpredictably, due to non-linear dynamics caused by crystallization and outgassing of the highly viscous lava in the dome's conduit. Domes undergo various processes such as growth, collapse, solidification and erosion.

Lava domes grow by endogenic dome growth or exogenic dome growth. The former implies dome interior expansion to accommodate new lava and the latter refers to superficial piling up of lava. It is the high viscosity of the lava that prevents it from flowing far from the vent from which it extrudes, creating a dome-like shape of sticky lava that then cools slowly in situ. Domes may reach heights of several hundred meters, and can grow slowly and steadily for months (e.g. Unzen volcano), years (e.g. Soufrière Hills volcano), or even centuries (e.g. Mount Merapi volcano). The sides of these structures are composed of unstable rock debris. Due to the intermittent buildup of gas pressure, erupting domes can often experience episodes of explosive eruption over time. If part of a lava dome collapses while it is still molten, it can produce pyroclastic flows, one of the most lethal forms of volcanic event. Other hazards associated with lava domes are the destruction of property, forest fires, and lahars triggered by pyroclastic flows near mud, snow and ice. Lava domes are one of the principal structural features of many stratovolcanoes worldwide. Lava domes are prone to unusually dangerous explosions since they contain rhyolitic silica-rich lava.

Characteristics of lava dome eruptions include shallow, long-period and hybrid seismicity, which is attributed to excess fluid pressures in the contributing vent chamber. Other characteristics of lava domes include their hemispherical dome shape, cycles of dome growth over long periods, and sudden onsets of violent explosive activity. The average rate of dome growth may be used as a rough indicator of magma supply, but it shows no systematic relationship to the timing or characteristics of lava dome explosions.

Related Landforms

Photo showing the bulging cryptodome of Mt. St. Helens on April 27, 1980.

Cryptodomes

A cryptodome (*kryptos*, "hidden, secret") is a dome-shaped structure created by accumulation of viscous magma at a shallow depth. One example of a cryptodome was in the May 1980 eruption of Mount St. Helens, where the explosive eruption began after a landslide caused the side of the volcano to fall, leading to explosive decompression of the subterranean cryptodome.

Lava Coulées

Coulées (or coulees) are lava domes that have experienced some flow away from their original position, thus resembling both lava domes and lava flows.

Chao dacite coulée flow-domes (left center), northern Chile, viewed from Landsat 8.

The world's largest known dacite flow is the Chao dacite dome complex, a huge coulée flow-dome between two volcanoes in northern Chile. This flow is over 14 kilometres (8.7 mi) long, has obvious flow features like pressure ridges, and a flow front 400 metres (1,300 ft) tall (the dark scalloped line at lower left). There is another prominent coulée flow on the flank of Llullaillaco volcano, in Argentina, and other examples in the Andes.

Volcanic Cone

Mayon Volcano in the Philippines has a symmetrical volcanic cone

Volcanic cones are among the simplest volcanic landforms. They are built by ejecta from a volcanic vent, piling up around the vent in the shape of a cone with a central crater. Volcanic cones are of different types, depending upon the nature and size of the fragments ejected during the eruption. Types of volcanic cones include stratocones, spatter cones, tuff cones, and cinder cones.

Stratocone

Stratocones are large cone-shaped volcanoes made up of lava flows, explosively erupted pyroclastic rocks, and igneous intrusives that are typically centered around a cylindrical vent. Unlike shield volcanoes, they are characterized by a steep profile and periodic, often alternating, explosive eruptions and effusive eruptions. Some have collapsed craters called calderas. The central core of a stratocone is commonly dominated by a central core of intrusive rocks that range from around 500 meters (1,600 ft) to over several kilometers in diameter. This central core is surrounded by multiple generations of lava flows, many of which are brecciated, and a wide range of pyroclastic rocks and reworked volcanic debris. The typical stratocone is an andesitic to dacitic volcano that is associated with subduction zones. They are also known as either *stratified volcano, composite cone, bedded volcano, cone of mixed type* or *Vesuvian-type volcano.*

Osorno volcano in Chile is an example of a well-developed stratocone.

Spatter Cone

Pu'u'Ō'ō, a cinder-and-spatter cone on Kīlauea, Hawaii

A spatter cone is a low, steep-sided hill or mound that consists of welded lava fragments, called *spatter,* which has formed around a lava fountain issuing from a central vent. Typically, spatter cones are about 3–5 meters (9.8–16.4 ft) high. In case of a linear fissure, lava fountaining will create broad embankments of spatter, called *spatter ramparts,* along both sides of the fissure. Spatter cones are more circular and cone shaped, while spatter ramparts are linear wall-like features.

Spatter cones and spatter ramparts are typically formed by lava fountaining associated with mafic, highly fluid lavas, such as those erupted in the Hawaiian Islands. As blobs of molten lava, *spatter,* are erupted into the air by a lava fountain, they can lack the time needed to cool completely before hitting the ground. Consequently, the spatter are not fully solid, like taffy, as they land and they bind to the underlying spatter as both often slowly ooze down the side of the cone. As a result, the spatter builds up a cone that is composed of spatter either agglutinated or welded to each other.

Tuff Cones (Ash Cones)

A tuff cone, sometimes called an ash cone, is a small monogenetic volcanic cone produced by phreatic (hydrovolcanic) explosions directly associated with magma brought to the surface through a conduit from a deep-seated magma reservoir. They are characterized by high rims that have a maximum relief of 100–800 meters (330–2,620 ft) above the crater floor and steep slopes that are greater than 25 degrees. They typically have a rim to rim diameter of 300–5,000 meters (980–16,400 ft). A tuff cone consists typically of thick-bedded pyroclastic flow and surge deposits created by eruption-fed density currents and bomb-scoria beds derived from fallout from its eruption column. The tuffs composing a tuff cone have commonly been altered, palagonitized, by either its interaction with groundwater or when it was deposited warm and wet. The pyroclastic deposits of tuff cones differ from the pyroclastic deposits of spatter cones by their lack or paucity of lava spatter, smaller grain-size, and excellent bedding. Typically, but not always, tuff cones lack associated lava flows.

Koko Crater is 10,000-year-old tuff cone, youngest in the Honolulu Volcanic Series

A **tuff ring** is a related type of small monogenetic volcano that is also produced by phreatic (hydrovolcanic) explosions directly associated with magma brought to the surface through a conduit from a deep-seated magma reservoir. . They are characterized by rims that have a low, broad topograhic profiles and gentle topographic slopes that are 25 degrees or less. The maximum thickness of the pyroclastic debris comprising the rim of a typical tuff ring is generally thin, less than 50 meters (160 ft) to 100 meters (330 ft) thick. The pyroclastic materials that comprise their rim consist primarily of relatively fresh and unaltered, distinctly and thin-bedded volcanic surge and air fall deposits. Their rims also can contain variable amounts of local country rock (bedrock) blasted out of their crater. In contrast to tuff cones, the crater of a tuff ring generally has been excavated below the existing ground surface. As a result, water commonly fills a tuff ring's crater to form a lake once eruptions cease.

Both tuff cones and their associated tuff rings were created by explosive eruptions from a vent where the magma is interacting with either groundwater or a shallow body of water as found within a lake or sea. The interaction between the magma, expanding steam, and volcanic gases resulted in the production and ejection of fine-grained pyroclastic debris called *ash* with the consistency of flour. The volcanic ash comprising a tuff cone accumulated either as fallout from eruption columns, from low-density volcanic surges and pyroclastic flows, or combination of these. Tuff cones are typically associated with volcanic eruptions within shallow bodies of water and tuff rings are associated with eruptions within either water saturated sediments and bedrock or permafrost

Next to spatter (scoria) cones, tuff cones and their associated tuff rings are among the most common types of volcanoes on Earth. An example of a tuff cone is Diamond Head at Waikīkī in Hawai□i. Clusters of pitted cones observed in the Nephentes/Amenthes region of Mars at the southern margin of the ancient Utopia impact basin are currently interpreted as being tuff cones and rings.

Cinder Cone

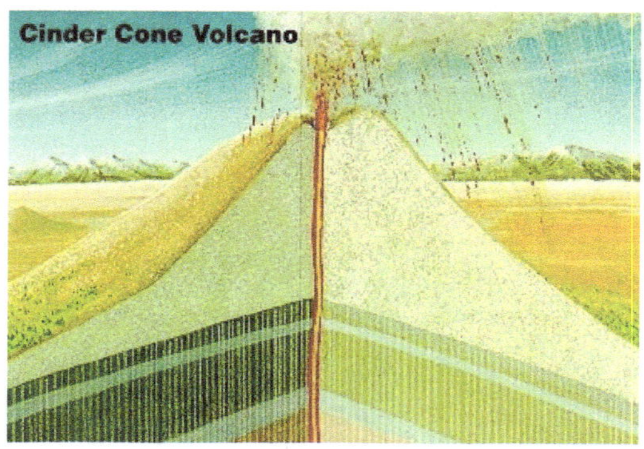

Cinder cone

Paricutin is a large cinder cone in Mexico.

Cinder cones, also known as scoria cones and less commonly scoria mounds, are small, steep-sided volcanic cones built of loose pyroclastic fragments, such as either volcanic clinkers, cinders, volcanic ash, or scoria. They consist of loose pyroclastic debris formed by explosive eruptions or lava fountains from a single, typically cylindrical, vent. As the gas-charged lava is blown violently into the air, it breaks into small fragments that solidify and fall as either cinders, clinkers, or scoria around the vent to form a cone that often is beautifully symmetrical; with slopes between 30-40°; and a nearly circular ground plan. Most cinder cones have a bowl-shaped crater at the summit. The basal diameters of cinder cones average about 800 meters (2,600 ft) and range from 250–2,500 meters (820–8,200 ft). The diameter of their craters range between 50–600 meters (160–1,970 ft). Cinder cones rarely rise more than 50–350 meters (160–1,150 ft) or so above their surroundings.

Cinder cones most commonly occur as isolated cones in large basaltic volcanic fields. In addition, they also occur in nested clusters in association with complex tuff ring and maar complexes. Finally, they are also common as parasitic and monogenetic cones on complex shield and stratovolcanoes. Globally, cinder cones are the most typical volcanic landform found within continental intraplate volcanic fields and also occur in some subduction zone settings as well. Parícutin, the Mexican cinder cone which was born in a cornfield on February 20, 1943, and Sunset Crater in Northern Arizona in the US Southwest are classic examples of cinder cones, as are ancient volcanic cones found in New Mexico's Petroglyph National Monument. Cone-shaped hills observed in satellite imagery of the calderas and volcanic cones of Ulysses Patera, Ulysses Colles and Hydraotes Chaos. are argued to be cinder cones.

Cinder cones typically only erupt once like Paricutin. As a result, they are considered to be monogenetic volcanoes and most of them form monogenetic volcanic fields. Cinder cones are typically active for very brief periods of time before becoming inactive. Their eruptions range in duration from a few days to a few years. Of observed cinder cone eruptions, 50% have lasted for less than 30 days, and 95% stopped within one year. In case of Paricutin in Mexico, its eruption lasted for nine years from 1943 to 1952. Rarely do they erupt either two, three, or more times. Later eruptions typically produce new cones within a volcanic field at separation distances of a few kilometers and separate by periods of 100 to 1,000 years. Within a volcanic field, eruptions can occur over a period of a million years. Once eruptions cease, being unconsolidated, cinder cones tend to erode rapidly unless further eruptions occur.

Rootless Cones

Rootless cones, also called pseudocraters, are volcanic cones that are not directly associated with a conduit that brought magma to the surface from a deep-seated magma reservoir. Generally, three types of rootless cones, littoral cones, explosion craters, and hornitos are recognized. Littoral cones and explosion craters are the result of mild explosions that were generated locally by the interaction of either hot lava or pyroclastic flows with water. Littoral cones typically form on the surface of a basaltic lava flow where it has entered into a body of water, usually a sea or ocean. Explosion craters form where either hot lava or pyroclastic flows have covered either marshy ground or water-saturated ground of some sort. Hornitos are rootless cones that are composed welded lava fragments and were formed on the surface of basaltic lava flows by the escape of gas and clots of molten lava through cracks or other openings in the crust of a lava flow.

References

- Poldervaart, A (1971). "Volcanicity and forms of extrusive bodies". In Green, J; Short, NM. Volcanic Landforms and Surface Features: A Photographic Atlas and Glossary. New York: Springer-Verlag. pp. 1–18. ISBN 978364265152-6.
- Cas, R.A.F., and J.V. Wright (1987) Volcanic Successions: Modern and Ancient, 1st ed. Chapman & Hall, London, United Kingdom. pp. 528 ISBN 978-0412446405
- Brož, P.; Hauber, E. (2013). "Hydrovolcanic tuff rings and cones as indicators for phreatomagmatic explosive eruptions on Mars". Journal of Geophysical Research: Planets. 118 (8): 1656–1675. doi:10.1002/jgre.20120.
- Brož, P.; Hauber, E. (2012). "A unique volcanic field in Tharsis, Mars: Pyroclastic cones as evidence for explosive eruptions". Icarus. 218 (1): 88–99. Bibcode:2012Icar..218...88B. doi:10.1016/j.icarus.2011.11.030.

5

Types of Volcanic Rocks

Sedimentary rock is one of the main types of rocks, which covers metamorphic rocks and igneous rocks. Sedimentary rocks form the deposition and cementation of the Earth's surface while the cooling and then solidification of magma or lava forms the igneous rocks. The topics discussed in the chapter are of great importance to broaden the existing knowledge on volcanic rocks.

Volcanic Rock

Volcanic rock (often shortened to volcanics in scientific contexts) is a rock formed from magma erupted from a volcano. In other words, it differs from other igneous rock by being of volcanic origin. Like all rock types, the concept of volcanic rock is artificial, and in nature volcanic rocks grade into hypabyssal and metamorphic rocks and constitute an important element of some sediments and sedimentary rocks. For these reasons, in geology, volcanics and shallow hypabyssal rocks are not always treated as distinct. In the context of Precambrian shield geology, the term "volcanic" is often applied to what are strictly metavolcanic rocks.

Ignimbrite is a deposit of a pyroclastic flow.

Volcanic rocks are among the most common rock types on Earth's surface, particularly in the oceans. On land, they are very common at plate boundaries and in flood basalt provinces.

Characteristics

Setting and Size

- Lava

- Tephra
- Volcanic bomb
- Lapilli
- Volcanic ash

Texture

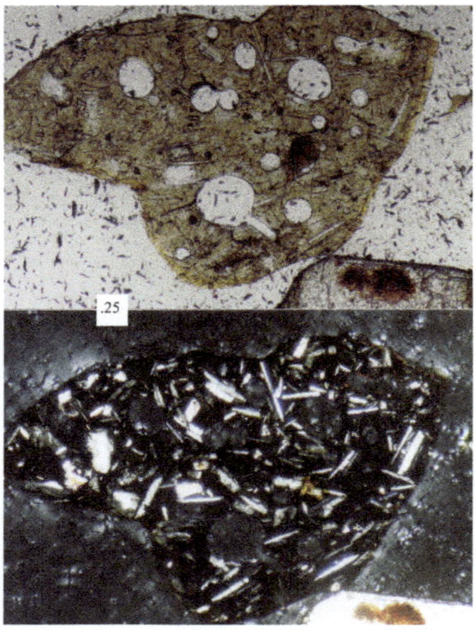

Photomicrograph of a volcanic lithic fragment (sand grain); upper picture is plane-polarized light, bottom picture is cross-polarized light, scale box at left-center is 0.25 millimeter.

Volcanic rocks are usually fine-grained or aphanitic to glass in texture. They often contain clasts of other rocks and phenocrysts. Phenocrysts are crystals that are larger than the matrix and are identifiable with the unaided eye. Rhomb porphyry is an example with large rhomb shaped phenocrysts embedded in a very fine grained matrix.

Volcanic rocks often have a vesicular texture caused by voids left by volatiles trapped in the molten lava. Pumice is a highly vesicular rock produced in explosive volcanic eruptions.

Chemistry

Most modern petrologists classify igneous rocks, including volcanic rocks, by their chemistry when dealing with their origin. The fact that different mineralogies and textures may be developed from the same initial magmas has led petrologists to rely heavily on chemistry to look at a volcanic rock's origin.

The chemistry of volcanic rocks is dependent on two things: the initial composition of the primary magma and the subsequent differentiation. Differentiation of most volcanic rocks tends to increase the silica (SiO_2) content, mainly by crystal fractionation.

The initial composition of most volcanic rocks is basaltic, albeit small differences in initial compositions may result in multiple differentiation series. The most common of these series are tholeiitic, calc-alkaline, and alkaline.

Mineralogy

Most volcanic rocks share a number of common minerals. Differentiation of volcanic rocks tends to increase the silica (SiO_2) content mainly by fractional crystallization. Thus, more evolved volcanic rocks tend to be richer in minerals with a higher amount if silica such as phyllo and tectosilicates including the feldspars, quartz polymorphs and muscovite. While still dominated by silicates, more primitive volcanic rocks have mineral assemblages with less silica, such as olivine and the pyroxenes. Bowen's reaction series correctly predicts the order of formation of the most common minerals in volcanic rocks.

Occasionally, a magma may pick up crystals that crystallized from another magma; these crystals are called xenocrysts. Diamonds found in kimberlites are rare but well-known xenocrysts; the kimberlites do not create the diamonds, but pick them up and transport them to the surface of the Earth.

Naming

Volcanic rocks are named according to both their chemical composition and texture. Basalt is a very common volcanic rock with low silica content. Rhyolite is a volcanic rock with high silica content. Rhyolite has silica content similar to that of granite while basalt is compositionally equal to gabbro. Intermediate volcanic rocks include andesite, dacite, trachyte, and latite.

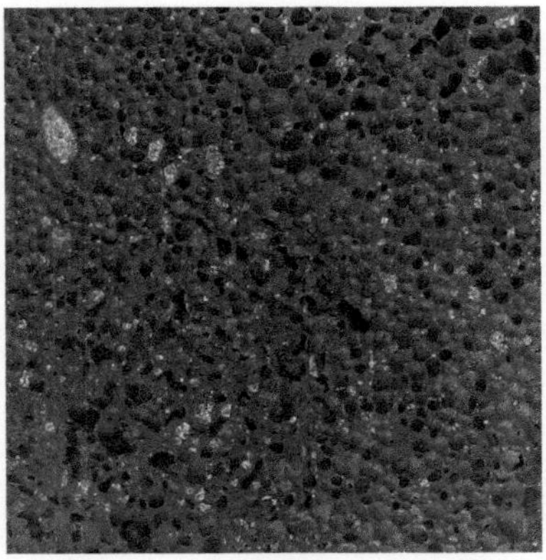

Vesicular olivine basalt from La Palma (green phenocrysts are olivine).

Pyroclastic rocks are the product of explosive volcanism. They are often felsic (high in silica). Pyroclastic rocks are often the result of volcanic debris, such as ash, bombs and tephra, and other volcanic ejecta. Examples of pyroclastic rocks are tuff and ignimbrite.

Shallow intrusions, which possess structure similar to volcanic rather than plutonic rocks are also considered to be volcanic.

Composition of Volcanic Rocks

'A'ā next to pāhoehoe lava at the Craters of the Moon National Monument and Preserve, Idaho, United States.

A German example of latite, a type of volcanic rock

The sub-family of rocks that form from volcanic lava are called igneous volcanic rocks (to differentiate them from igneous rocks that form from magma below the surface, called igneous plutonic rocks).

The lavas of different volcanoes, when cooled and hardened, differ much in their appearance and composition. If a rhyolite lava-stream cools quickly, it can quickly freeze into a black glassy substance called obsidian. When filled with bubbles of gas, the same lava may form the spongy appearing pumice. Allowed to cool slowly, it forms a light-colored, uniformly solid rock called rhyolite.

A sample of rhyolite

Basaltic scoria from Amsterdam Island in the Indian Ocean

The lavas, having cooled rapidly in contact with the air or water, are mostly finely crystalline or have at least fine-grained ground-mass representing that part of the viscous semi-crystalline lava flow that was still liquid at the moment of eruption. At this time they were exposed only to atmospheric pressure, and the steam and other gases, which they contained in great quantity were free to escape; many important modifications arise from this, the most striking being the frequent presence of numerous steam cavities (vesicular structure) often drawn out to elongated shapes subsequently filled up with minerals by infiltration (amygdaloidal structure).

As crystallization was going on while the mass was still creeping forward under the surface of the Earth, the latest formed minerals (in the ground-mass) are commonly arranged in subparallel winding lines that follow the direction of movement (fluxion or fluidal structure)—and larger early minerals that previously crystallized may show the same arrangement. Most lavas fall considerably below their original temperatures before emitted. In their behavior, they present a close analogy to hot solutions of salts in water, which, when they approach the saturation temperature, first deposit a crop of large, well-formed crystals (labile stage) and subsequently precipitate clouds of smaller less perfect crystalline particles (metastable stage).

In igneous rocks the first generation of crystals generally forms before the lava has emerged to the surface, that is to say, during the ascent from the subterranean depths to the crater of the volcano. It has frequently been verified by observation that freshly emitted lavas contain large crystals borne along in a molten, liquid mass. The large, well-formed, early crystals (phenocrysts) are said to be porphyritic; the smaller crystals of the surrounding matrix or ground-mass belong to the post-effusion stage. More rarely lavas are completely fused at the moment of ejection; they may then cool to form a non-porphyritic, finely crystalline rock, or if more rapidly chilled may in large part be non-crystalline or glassy (vitreous rocks such as obsidian, tachylyte, pitchstone).

A common feature of glassy rocks is the presence of rounded bodies (spherulites), consisting of fine divergent fibres radiating from a center; they consist of imperfect crystals of feldspar, mixed with quartz or tridymite; similar bodies are often produced artificially in glasses that are allowed to cool slowly. Rarely these spherulites are hollow or consist of concentric shells with spaces between (lithophysae). Perlitic structure, also common in glasses, consists of the presence of concentric rounded cracks owing to contraction on cooling.

Volcanic rocks, Porto Moniz, Madeira.

The phenocrysts or porphyritic minerals are not only larger than those of the ground-mass; as the matrix was still liquid when they formed they were free to take perfect crystalline shapes, without interference by the pressure of adjacent crystals. They seem to have grown rapidly, as they are often filled with enclosures of glassy or finely crystalline material like that of the ground-mass . Microscopic examination of the phenocrysts often reveals that they have had a complex history. Very frequently they show layers of different composition, indicated by variations in color or other optical properties; thus augite may be green in the center surrounded by various shades of brown; or they may be pale green centrally and darker green with strong pleochroism (aegirine) at the periphery.

In the feldspars the center is usually richer in calcium than the surrounding layers, and successive zones may often be noted, each less calcic than those within it. Phenocrysts of quartz (and of other minerals), instead of sharp, perfect crystalline faces, may show rounded corroded surfaces, with the points blunted and irregular tongue-like projections of the matrix into the substance of the crystal. It is clear that after the mineral had crystallized it was partly again dissolved or corroded at some period before the matrix solidified.

Corroded phenocrysts of biotite and hornblende are very common in some lavas; they are surrounded by black rims of magnetite mixed with pale green augite. The hornblende or biotite substance has proved unstable at a certain stage of consolidation, and has been replaced by a paramorph of augite and magnetite, which may partially or completely substitute for the original crystal but still retains its characteristic outlines.

Igneous Rock

Igneous rock (derived from the Latin word *ignis* meaning fire) is one of the three main rock types, the others being sedimentary and metamorphic. Igneous rock is formed through the cooling and solidification of magma or lava. Igneous rock may form with or without crystallization, either below the surface as intrusive (plutonic) rocks or on the surface as extrusive (volcanic) rocks. This magma can be derived from partial melts of existing rocks in either a planet's mantle or crust. Typically, the melting is caused by one or more of three processes: an increase in temperature, a decrease in pressure, or a change in composition.

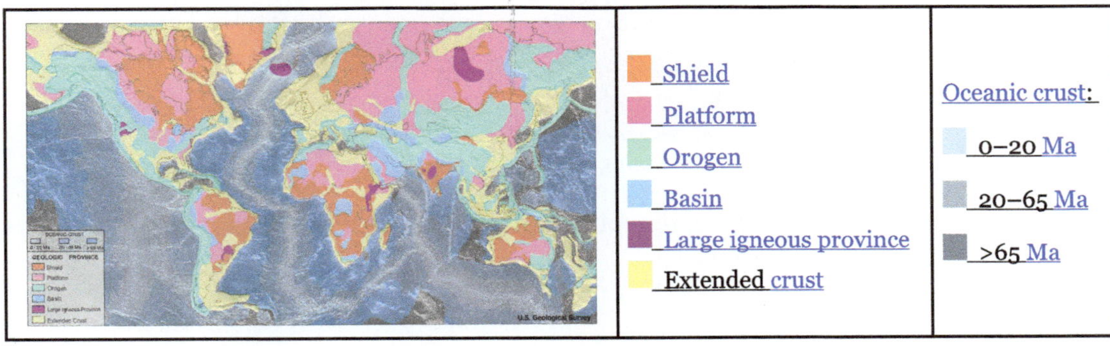

Geologic provinces of the World (USGS)

Geological Significance

Igneous and metamorphic rocks make up 90–95% of the top 16 km of the Earth's crust by volume.

Igneous rocks are geologically important because:

- their minerals and global chemistry give information about the composition of the mantle, from which some igneous rocks are extracted, and the temperature and pressure conditions that allowed this extraction, and/or of other pre-existing rock that melted;
- their absolute ages can be obtained from various forms of radiometric dating and thus can be compared to adjacent geological strata, allowing a time sequence of events;
- their features are usually characteristic of a specific tectonic environment, allowing tectonic reconstitutions;
- in some special circumstances they host important mineral deposits (ores): for example, tungsten, tin, and uranium are commonly associated with granites and diorites, whereas ores of chromium and platinum are commonly associated with gabbros.

Morphology and Setting

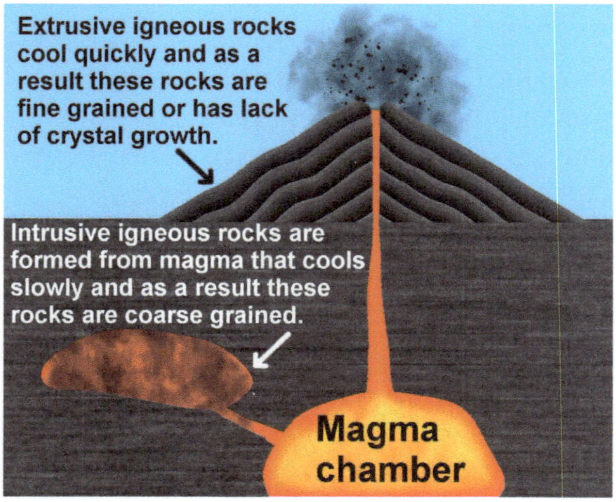

Forming of igneous rock

In terms of modes of occurrence, igneous rocks can be either intrusive (plutonic), extrusive (volcanic) or hypabyssal.

Intrusive

Intrusive igneous rocks are formed from magma that cools and solidifies within the crust of a planet, surrounded by pre-existing rock (called country rock); the magma cools slowly and, as a result, these rocks are coarse grained. The mineral grains in such rocks can generally be identified with the naked eye. Intrusive rocks can also be classified according to the shape and size of the intrusive body and its relation to the other formations into which it intrudes. Typical intrusive formations are batholiths, stocks, laccoliths, sills and dikes. When the magma solidifies within the earth's crust, it cools slowly forming coarse textured rocks, such as granite, gabbro, or diorite.

Close-up of granite (an intrusive igneous rock) exposed in Chennai, India.

The central cores of major mountain ranges consist of intrusive igneous rocks, usually granite. When exposed by erosion, these cores (called *batholiths*) may occupy huge areas of the Earth's surface.

Coarse grained intrusive igneous rocks that form at depth within the crust are termed as *abyssal*; intrusive igneous rocks that form near the surface are termed *hypabyssal.*

Extrusive

Sample of basalt (an extrusive igneous rock), found in Massachusetts

Extrusive igneous rocks, also known as volcanic rocks, are formed at the crust's surface as a result of the partial melting of rocks within the mantle and crust. Extrusive igneous rocks cool and solidify quicker than intrusive igneous rocks. They are formed by the cooling of molten magma on the earth's surface. The magma, which is brought to the surface through fissures or volcanic eruptions, solidifies at a faster rate. Hence such rocks are smooth, crystalline and fine grained. Basalt is a common extrusive igneous rock and forms lava flows, lava sheets and lava plateaus. Some kinds of basalt solidify to form long polygonal columns. The Giant's Causeway in Antrim, Northern Ireland is an example.

The molten rock, with or without suspended crystals and gas bubbles, is called magma. It rises because it is less dense than the rock from which it was created. When magma reaches the surface from beneath water or air, it is called lava. Eruptions of volcanoes into air are termed *subaerial,* whereas those occurring underneath the ocean are termed *submarine.* Black smokers and mid-ocean ridge basalt are examples of submarine volcanic activity.

The volume of extrusive rock erupted annually by volcanoes varies with plate tectonic setting. Extrusive rock is produced in the following proportions:

- divergent boundary: 73%
- convergent boundary (subduction zone): 15%
- hotspot: 12%.

Magma that erupts from a volcano behaves according to its viscosity, determined by temperature, composition, and crystal content. High-temperature magma, most of which is basaltic in composition, behaves in a manner similar to thick oil and, as it cools, treacle. Long, thin basalt flows with pahoehoe surfaces are common. Intermediate composition magma, such as andesite, tends to form cinder cones of intermingled ash, tuff and lava, and may have a viscosity similar to thick, cold molasses or even rubber when erupted. Felsic magma, such as rhyolite, is usually erupted at low temperature and is up to 10,000 times as viscous as basalt. Volcanoes with rhyolitic magma commonly erupt explosively, and rhyolitic lava flows are typically of limited extent and have steep margins, because the magma is so viscous.

Felsic and intermediate magmas that erupt often do so violently, with explosions driven by the release of dissolved gases—typically water vapour, but also carbon dioxide. Explosively erupted pyroclastic material is called tephra and includes tuff, agglomerate and ignimbrite. Fine volcanic ash is also erupted and forms ash tuff deposits, which can often cover vast areas.

Because lava cools and crystallizes rapidly, it is fine grained. If the cooling has been so rapid as to prevent the formation of even small crystals after extrusion, the resulting rock may be mostly glass (such as the rock obsidian). If the cooling of the lava happened more slowly, the rocks would be coarse-grained.

Because the minerals are mostly fine-grained, it is much more difficult to distinguish between the different types of extrusive igneous rocks than between different types of intrusive igneous rocks. Generally, the mineral constituents of fine-grained extrusive igneous rocks can only be determined by examination of thin sections of the rock under a microscope, so only an approximate classification can usually be made in the field.

Hypabyssal

Hypabyssal igneous rocks are formed at a depth in between the plutonic and volcanic rocks. These are formed due to cooling and resultant solidification of rising magma just beneath the earth surface. Hypabyssal rocks are less common than plutonic or volcanic rocks and often form dikes, sills, laccoliths, lopoliths, or phacoliths.

Classification

Igneous rocks are classified according to mode of occurrence, texture, mineralogy, chemical composition, and the geometry of the igneous body.

The classification of the many types of different igneous rocks can provide us with important information about the conditions under which they formed. Two important variables used for the classification of igneous rocks are particle size, which largely depends on the cooling history, and the mineral composition of the rock. Feldspars, quartz or feldspathoids, olivines, pyroxenes, amphiboles, and micas are all important minerals in the formation of almost all igneous rocks, and they are basic to the classification of these rocks. All other minerals present are regarded as nonessential in almost all igneous rocks and are called *accessory minerals*. Types of igneous rocks with other essential minerals are very rare, and these rare rocks include those with essential carbonates.

In a simplified classification, igneous rock types are separated on the basis of the type of feldspar present, the presence or absence of quartz, and in rocks with no feldspar or quartz, the type of iron or magnesium minerals present. Rocks containing quartz (silica in composition) are silica-oversaturated. Rocks with feldspathoids are silica-undersaturated, because feldspathoids cannot coexist in a stable association with quartz.

Igneous rocks that have crystals large enough to be seen by the naked eye are called phaneritic; those with crystals too small to be seen are called aphanitic. Generally speaking, phaneritic implies an intrusive origin; aphanitic an extrusive one.

An igneous rock with larger, clearly discernible crystals embedded in a finer-grained matrix is termed porphyry. Porphyritic texture develops when some of the crystals grow to considerable size before the main mass of the magma crystallizes as finer-grained, uniform material.

Igneous rocks are classified on the basis of texture and composition. Texture refers to the size, shape, and arrangement of the mineral grains or crystals of which the rock is composed.

Texture

Gabbro specimen showing phaneritic texture; Rock Creek Canyon, eastern Sierra Nevada, California; scale bar is 2.0 cm.

Texture is an important criterion for the naming of volcanic rocks. The texture of volcanic rocks, including the size, shape, orientation, and distribution of mineral grains and the intergrain relationships, will determine whether the rock is termed a tuff, a pyroclastic lava or a simple lava.

However, the texture is only a subordinate part of classifying volcanic rocks, as most often there needs to be chemical information gleaned from rocks with extremely fine-grained groundmass or from airfall tuffs, which may be formed from volcanic ash.

Textural criteria are less critical in classifying intrusive rocks where the majority of minerals will be visible to the naked eye or at least using a hand lens, magnifying glass or microscope. Plutonic rocks also tend to be less texturally varied and less prone to gaining structural fabrics. Textural terms can be used to differentiate different intrusive phases of large plutons, for instance porphyritic margins to large intrusive bodies, porphyry stocks and subvolcanic dikes (apophyses). Mineralogical classification is most often used to classify plutonic rocks. Chemical classifications are preferred to classify volcanic rocks, with phenocryst species used as a prefix, e.g. "olivine-bearing picrite" or "orthoclase-phyric rhyolite".

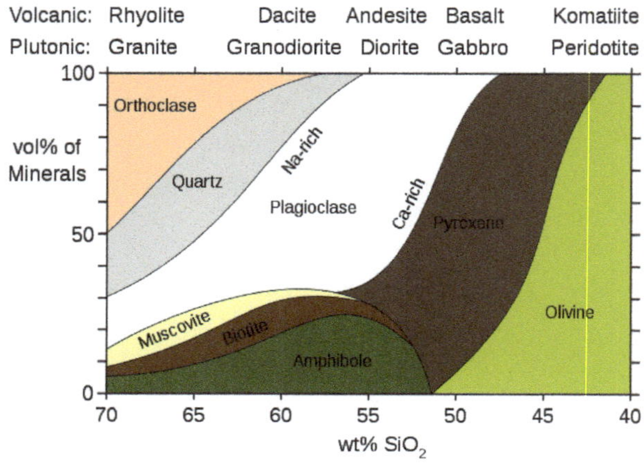

Basic classification scheme for igneous rocks on their mineralogy. If the approximate volume fractions of minerals in the rock are known, the rock name and silica content can be read off the diagram. This is not an exact method, because the classification of igneous rocks also depends on other components than silica, yet in most cases it is a good first guess.

Chemical classification and Petrology

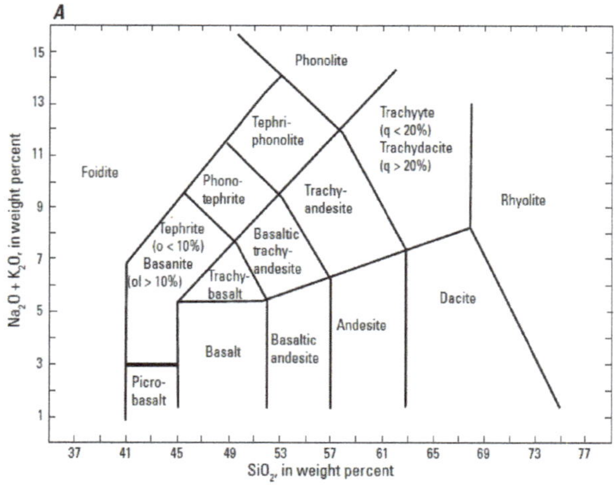

Total alkali versus silica classification scheme (TAS) as proposed in Le Maitre's 2002 Igneous Rocks - A classification and glossary of terms

Igneous rocks can be classified according to chemical or mineralogical parameters.

Chemical: total alkali-silica content (TAS diagram) for volcanic rock classification used when modal or mineralogic data is unavailable:

- *felsic* igneous rocks containing a high silica content, greater than 63% SiO_2 (examples granite and rhyolite)
- *intermediate* igneous rocks containing between 52 – 63% SiO_2 (example andesite and dacite)
- *mafic* igneous rocks have low silica 45 – 52% and typically high iron – magnesium content (example gabbro and basalt)
- *ultramafic rock* igneous rocks with less than 45% silica. (examples picrite, komatiite and peridotite)
- *alkalic* igneous rocks with 5 – 15% alkali ($K_2O + Na_2O$) content or with a molar ratio of alkali to silica greater than 1:6. (examples phonolite and trachyte)

Chemical classification also extends to differentiating rocks that are chemically similar according to the TAS diagram, for instance;

- Ultrapotassic; rocks containing molar $K_2O/Na_2O > 3$
- Peralkaline; rocks containing molar $(K_2O + Na_2O)/ Al_2O_3 > 1$
- Peraluminous; rocks containing molar $(K_2O + Na_2O)/ Al_2O_3 < 1$

An idealized mineralogy (the normative mineralogy) can be calculated from the chemical composition, and the calculation is useful for rocks too fine-grained or too altered for identification of minerals that crystallized from the melt. For instance, normative quartz classifies a rock as silica-oversaturated; an example is rhyolite. In an older terminology, silica oversaturated rocks were called *silicic* or *acidic* where the SiO_2 was greater than 66% and the family term *quartzolite* was applied to the most silicic. A normative feldspathoid classifies a rock as silica-undersaturated; an example is nephelinite.

History of Classification

In 1902, a group of American petrographers proposed that all existing classifications of igneous rocks should be discarded and replaced by a "quantitative" classification based on chemical analysis. They showed how vague, and often unscientific, much of the existing terminology was and argued that as the chemical composition of an igneous rock was its most fundamental characteristic, it should be elevated to prime position.

Geological occurrence, structure, mineralogical constitution—the hitherto accepted criteria for the discrimination of rock species—were relegated to the background. The completed rock analysis is first to be interpreted in terms of the rock-forming minerals which might be expected to be formed when the magma crystallizes, e.g., quartz feldspars, olivine, akermannite, Feldspathoids, magnetite, corundum, and so on, and the rocks are divided into groups strictly according to the relative proportion of these minerals to one another.

Mineralogical Classification

For volcanic rocks, mineralogy is important in classifying and naming lavas. The most important criterion is the phenocryst species, followed by the groundmass mineralogy. Often, where the groundmass is aphanitic, chemical classification must be used to properly identify a volcanic rock.

Mineralogic Contents – Felsic Versus Mafic

- *felsic* rock, highest content of silicon, with predominance of quartz, alkali feldspar and/or feldspathoids: *the felsic minerals*; these rocks (e.g., granite, rhyolite) are usually light coloured, and have low density.
- *mafic* rock, lesser content of silicon relative to felsic rocks, with predominance of mafic minerals pyroxenes, olivines and calcic plagioclase; these rocks (example, basalt, gabbro) are usually dark coloured, and have a higher density than felsic rocks.
- *ultramafic* rock, lowest content of silicon, with more than 90% of mafic minerals (e.g., dunite).

For intrusive, plutonic and usually phaneritic igneous rocks (where all minerals are visible at least via microscope), the mineralogy is used to classify the rock. This usually occurs on ternary diagrams, where the relative proportions of three minerals are used to classify the rock.

The following table is a simple subdivision of igneous rocks according to both their composition and mode of occurrence.

	Composition			
Mode of occurrence	Felsic	Intermediate	Mafic	Ultramafic
Intrusive	Granite	Diorite	Gabbro	Peridotite
Extrusive	Rhyolite	Andesite	Basalt	Komatiite

	Essential rock forming silicates			
	Felsic	Intermediate	Mafic	Ultramafic
Coarse Grained	Granite	Diorite	Gabbro	Peridotite
Medium Grained			Diabase	
Fine Grained	Rhyolite	Andesite	Basalt	Komatiite

Example of Classification

Granite is an igneous intrusive rock (crystallized at depth), with felsic composition (rich in silica and predominately quartz plus potassium-rich feldspar plus sodium-rich plagioclase) and phaneritic, subeuhedral texture (minerals are visible to the unaided eye and commonly some of them retain original crystallographic shapes).

Magma Origination

The Earth's crust averages about 35 kilometers thick under the continents, but averages only some 7–10 kilometers beneath the oceans. The continental crust is composed primarily of sedimentary rocks resting on a crystalline *basement* formed of a great variety of metamorphic and igneous rocks, including granulite and granite. Oceanic crust is composed primarily of basalt and gabbro. Both continental and oceanic crust rest on peridotite of the mantle.

Rocks may melt in response to a decrease in pressure, to a change in composition (such as an addition of water), to an increase in temperature, or to a combination of these processes.

Other mechanisms, such as melting from a meteorite impact, are less important today, but impacts during the accretion of the Earth led to extensive melting, and the outer several hundred kilometers of our early Earth was probably an ocean of magma. Impacts of large meteorites in the last few hundred million years have been proposed as one mechanism responsible for the extensive basalt magmatism of several large igneous provinces.

Decompression

Decompression melting occurs because of a decrease in pressure.

The solidus temperatures of most rocks (the temperatures below which they are completely solid) increase with increasing pressure in the absence of water. Peridotite at depth in the Earth's mantle may be hotter than its solidus temperature at some shallower level. If such rock rises during the convection of solid mantle, it will cool slightly as it expands in an adiabatic process, but the cooling is only about 0.3 °C per kilometer. Experimental studies of appropriate peridotite samples document that the solidus temperatures increase by 3 °C to 4 °C per kilometer. If the rock rises far enough, it will begin to melt. Melt droplets can coalesce into larger volumes and be intruded upwards. This process of melting from the upward movement of solid mantle is critical in the evolution of the Earth.

Decompression melting creates the ocean crust at mid-ocean ridges. It also causes volcanism in intraplate regions, such as Europe, Africa and the Pacific sea floor. There, it is variously attributed either to the rise of mantle plumes (the "Plume hypothesis") or to intraplate extension (the "Plate hypothesis").

Effects of water and Carbon Dioxide

The change of rock composition most responsible for the creation of magma is the addition of water. Water lowers the solidus temperature of rocks at a given pressure. For example, at a depth of about 100 kilometers, peridotite begins to melt near 800 °C in the presence of excess water, but near or above about 1,500 °C in the absence of water. Water is driven out of the oceanic lithosphere in subduction zones, and it causes melting in the overlying mantle. Hydrous magmas composed of basalt and andesite are produced directly and indirectly as results of dehydration during the subduction process. Such magmas, and those derived from them, build up island arcs such as those in the Pacific Ring of Fire. These magmas form rocks of the calc-alkaline series, an important part of the continental crust.

The addition of carbon dioxide is relatively a much less important cause of magma formation than the addition of water, but genesis of some silica-undersaturated magmas has been attributed to the dominance of carbon dioxide over water in their mantle source regions. In the presence of carbon dioxide, experiments document that the peridotite solidus temperature decreases by about 200 °C in a narrow pressure interval at pressures corresponding to a depth of about 70 km. At greater depths, carbon dioxide can have more effect: at depths to about 200 km, the temperatures of initial melting of a carbonated peridotite composition were determined to be 450 °C to 600 °C lower than for the same composition with no carbon dioxide. Magmas of rock types such as nephelinite, carbonatite, and kimberlite are among those that may be generated following an influx of carbon dioxide into mantle at depths greater than about 70 km.

Temperature Increase

Increase in temperature is the most typical mechanism for formation of magma within continental crust. Such temperature increases can occur because of the upward intrusion of magma from the mantle. Temperatures can also exceed the solidus of a crustal rock in continental crust thickened by compression at a plate boundary. The plate boundary between the Indian and Asian continental masses provides a well-studied example, as the Tibetan Plateau just north of the boundary has crust about 80 kilometers thick, roughly twice the thickness of normal continental crust. Studies of electrical resistivity deduced from magnetotelluric data have detected a layer that appears to contain silicate melt and that stretches for at least 1,000 kilometers within the middle crust along the southern margin of the Tibetan Plateau. Granite and rhyolite are types of igneous rock commonly interpreted as products of the melting of continental crust because of increases in temperature. Temperature increases also may contribute to the melting of lithosphere dragged down in a subduction zone.

Magma Evolution

Most magmas only entirely melt for small parts of their histories. More typically, they are mixes of melt and crystals, and sometimes also of gas bubbles. Melt, crystals, and bubbles usually have different densities, and so they can separate as magmas evolve.

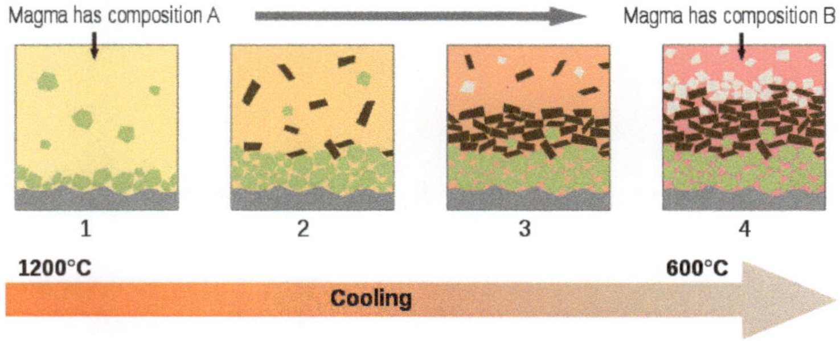

Schematic diagrams showing the principles behind fractional crystallisation in a magma. While cooling, the magma evolves in composition because different minerals crystallize from the melt. **1**: olivine crystallizes; **2**: olivine and pyroxene crystallize; **3**: pyroxene and plagioclase crystallize; **4**: plagioclase crystallizes. At the bottom of the magma reservoir, a cumulate rock forms.

As magma cools, minerals typically crystallize from the melt at different temperatures (fractional crystallization). As minerals crystallize, the composition of the residual melt typically changes. If crystals separate from the melt, then the residual melt will differ in composition from the parent magma. For instance, a magma of gabbroic composition can produce a residual melt of granitic composition if early formed crystals are separated from the magma. Gabbro may have a liquidus temperature near 1,200 °C, and the derivative granite-composition melt may have a liquidus temperature as low as about 700 °C. Incompatible elements are concentrated in the last residues of magma during fractional crystallization and in the first melts produced during partial melting: either process can form the magma that crystallizes to pegmatite, a rock type commonly enriched in incompatible elements. Bowen's reaction series is important for understanding the idealised sequence of fractional crystallisation of a magma.

Magma composition can be determined by processes other than partial melting and fractional crystallization. For instance, magmas commonly interact with rocks they intrude, both by melting those rocks and by reacting with them. Magmas of different compositions can mix with one another. In rare cases, melts can separate into two immiscible melts of contrasting compositions.

There are relatively few minerals that are important in the formation of common igneous rocks, because the magma from which the minerals crystallize is rich in only certain elements: silicon, oxygen, aluminium, sodium, potassium, calcium, iron, and magnesium. These are the elements that combine to form the silicate minerals, which account for over ninety percent of all igneous rocks. The chemistry of igneous rocks is expressed differently for major and minor elements and for trace elements. Contents of major and minor elements are conventionally expressed as weight percent oxides (e.g., 51% SiO_2, and 1.50% TiO_2). Abundances of trace elements are conventionally expressed as parts per million by weight (e.g., 420 ppm Ni, and 5.1 ppm Sm). The term "trace element" is typically used for elements present in most rocks at abundances less than 100 ppm or so, but some trace elements may be present in some rocks at abundances exceeding 1,000 ppm. The diversity of rock compositions has been defined by a huge mass of analytical data—over 230,000 rock analyses can be accessed on the web through a site sponsored by the U. S. National Science Foundation.

Etymology

The word "igneous" is derived from the Latin *ignis*, meaning "of fire". Volcanic rocks are named after Vulcan, the Roman name for the god of fire. Intrusive rocks are also called "plutonic" rocks, named after Pluto, the Roman god of the underworld.

Sedimentary Rock

Sedimentary rocks are types of rock that are formed by the deposition and subsequent cementation of that material at the Earth's surface and within bodies of water. Sedimentation is the collective name for processes that cause mineral and/or organic particles (detritus) to settle in place. The particles that form a sedimentary rock by accumulating are called sediment. Before being deposited, the sediment was formed by weathering and erosion from the source area, and then transported to the place of deposition by water, wind, ice, mass movement or glaciers, which are called agents of denudation. Sedimentation may also occur as minerals precipitate from water solution or shells of aquatic creatures settle out of suspension.

Middle Triassic marginal marine sequence of siltstones (reddish layers at the cliff base) and limestones (brown rocks above), Virgin Formation, southwestern Utah, USA

Sedimentary rocks on Mars, investigated by NASA's Curiosity Mars rover

Steeply dipping sedimentary rock strata along the Chalous Road in northern Iran

The sedimentary rock cover of the continents of the Earth's crust is extensive, but the total contribution of sedimentary rocks is estimated to be only 8% of the total volume of the crust. Sedimentary rocks are only a thin veneer over a crust consisting mainly of igneous and metamorphic rocks. Sedimentary rocks are deposited in layers as strata, forming a structure called bedding. The study of sedimentary rocks and rock strata provides information about the subsurface that is useful for civil engineering, for example in the construction of roads, houses, tunnels, canals or other structures. Sedimentary rocks are also important sources of natural resources like coal, fossil fuels, drinking water or ores.

The study of the sequence of sedimentary rock strata is the main source for an understanding of the Earth's history, including palaeogeography, paleoclimatology and the history of life. The scientific discipline that studies the properties and origin of sedimentary rocks is called sedimentology. Sedimentology is part of both geology and physical geography and overlaps partly with other disciplines in the Earth sciences, such as pedology, geomorphology, geochemistry and structural geology. Sedimentary rocks have also been found on Mars.

Classification based on Origin

Sedimentary rocks can be subdivided into four groups based on the processes responsible for their formation: clastic sedimentary rocks, biochemical (biogenic) sedimentary rocks, chemical sedimentary rocks, and a fourth category for "other" sedimentary rocks formed by impacts, volcanism, and other minor processes.

Clastic Sedimentary Rocks

Clastic sedimentary rocks are composed of other rock fragments that were cemented by silicate minerals. Clastic rocks are composed largely of quartz, feldspar, rock (lithic) fragments, clay minerals, and mica; any type of mineral may be present, but they in general represent the minerals that exist locally.

Claystone deposited in Glacial Lake Missoula, Montana, United States. Note the very fine and flat bedding, common for distal lacustrine deposition.

Clastic sedimentary rocks, are subdivided according to the dominant particle size. Most geologists use the Udden-Wentworth grain size scale and divide unconsolidated sediment into three fractions: gravel (>2 mm diameter), sand (1/16 to 2 mm diameter), and mud (clay is <1/256 mm and silt is between 1/16 and 1/256 mm). The classification of clastic sedimentary rocks parallels this scheme; conglomerates and breccias are made mostly of gravel, sandstones are made mostly of sand, and mudrocks are made mostly of the finest material. This tripartite subdivision is mirrored by the broad categories of rudites, arenites, and lutites, respectively, in older literature.

The subdivision of these three broad categories is based on differences in clast shape:- conglomerates and breccias), composition (sandstones), grain size and/or texture (mudrocks).

Conglomerates and Breccias

Conglomerates are dominantly composed of rounded gravel, while breccias are composed of dominantly angular gravel.

Sandstones

Sedimentary rock with sandstone in Malta

Sandstone classification schemes vary widely, but most geologists have adopted the Dott scheme, which uses the relative abundance of quartz, feldspar, and lithic framework grains and the abundance of a muddy matrix between the larger grains.

Composition of framework grains

The relative abundance of sand-sized framework grains determines the first word in a sandstone name. Naming depends on the dominance of the three most abundant components quartz, feldspar, or the lithic fragments that originated from other rocks. All other minerals are considered accessories and not used in the naming of the rock, regardless of abundance.

- Quartz sandstones have >90% quartz grains
- Feldspathic sandstones have <90% quartz grains and more feldspar grains than lithic grains
- Lithic sandstones have <90% quartz grains and more lithic grains than feldspar grains

Abundance of muddy matrix material between sand grains

When sand-sized particles are deposited, the space between the grains either remains open or is filled with mud (silt and/or clay sized particle).

- "Clean" sandstones with open pore space (that may later be filled with matrix material) are called arenites.
- Muddy sandstones with abundant (>10%) muddy matrix are called wackes.

Six sandstone names are possible using the descriptors for grain composition (quartz-, feldspathic-, and lithic-) and the amount of matrix (wacke or arenite). For example, a quartz arenite would be composed of mostly (>90%) quartz grains and have little or no clayey matrix between the grains, a lithic wacke would have abundant lithic grains and abundant muddy matrix, etc.

Although the Dott classification scheme is widely used by sedimentologists, common names like greywacke, arkose, and quartz sandstone are still widely used by non-specialists and in popular literature.

Mudrocks

Lower Antelope Canyon was carved out of the surrounding sandstone by both mechanical weathering and chemical weathering. Wind, sand, and water from flash flooding are the primary weathering agents.

Mudrocks are sedimentary rocks composed of at least 50% silt- and clay-sized particles. These relatively fine-grained particles are commonly transported by turbulent flow in water or air, and deposited as the flow calms and the particles settle out of suspension.

Most authors presently use the term "mudrock" to refer to all rocks composed dominantly of mud. Mudrocks can be divided into siltstones, composed dominantly of silt-sized particles; mudstones with subequal mixture of silt- and clay-sized particles; and claystones, composed mostly of clay-sized particles. Most authors use "shale" as a term for a fissile mudrock (regardless of grain size) although some older literature uses the term "shale" as a synonym for mudrock.

Biochemical Sedimentary Rocks

Outcrop of Ordovician oil shale (kukersite), northern Estonia

Biochemical sedimentary rocks are created when organisms use materials dissolved in air or water to build their tissue. Examples include:

- Most types of limestone are formed from the calcareous skeletons of organisms such as corals, mollusks, and foraminifera.
- Coal, formed from plants that have removed carbon from the atmosphere and combined it with other elements to build their tissue.
- Deposits of chert formed from the accumulation of siliceous skeletons of microscopic organisms such as radiolaria and diatoms.

Chemical Sedimentary Rocks

Chemical sedimentary rock forms when mineral constituents in solution become supersaturated and inorganically precipitate. Common chemical sedimentary rocks include oolitic limestone and rocks composed of evaporite minerals, such as halite (rock salt), sylvite, barite and gypsum.

"Other" Sedimentary Rocks

This fourth miscellaneous category includes rocks formed by Pyroclastic flows, impact breccias, volcanic breccias, and other relatively uncommon processes.

Compositional Classification Schemes

Alternatively, sedimentary rocks can be subdivided into compositional groups based on their mineralogy:

- Siliciclastic sedimentary rocks, are dominantly composed of silicate minerals. The sediment that makes up these rocks was transported as bed load, suspended load, or by sediment gravity flows. Siliciclastic sedimentary rocks are subdivided into conglomerates and breccias, sandstone, and mudrocks.
- Carbonate sedimentary rocks are composed of calcite (rhombohedral CaCO3), aragonite (orthorhombic CaCO3), dolomite (CaMg(CO3)2), and other carbonate minerals based on the CO2−3 ion. Common examples include limestone and dolostone.
- Evaporite sedimentary rocks are composed of minerals formed from the evaporation of water. The most common evaporite minerals are carbonates (calcite and others based on CO2−3), chlorides (halite and others built on Cl−), and sulfates (gypsum and others built on SO2−4). Evaporite rocks commonly include abundant halite (rock salt), gypsum, and anhydrite.
- Organic-rich sedimentary rocks have significant amounts of organic material, generally in excess of 3% total organic carbon. Common examples include coal, oil shale as well as source rocks for oil and natural gas.
- Siliceous sedimentary rocks are almost entirely composed of silica (SiO 2), typically as chert, opal, chalcedony or other microcrystalline forms.
- Iron-rich sedimentary rocks are composed of >15% iron; the most common forms are banded iron formations and ironstones.
- Phosphatic sedimentary rocks are composed of phosphate minerals and contain more than 6.5% phosphorus; examples include deposits of phosphate nodules, bone beds, and phosphatic mudrocks.

Deposition and Transformation

Sediment Transport and Deposition

Sedimentary rocks are formed when sediment is deposited out of air, ice, wind, gravity, or water flows carrying the particles in suspension. This sediment is often formed when weathering and erosion break down a rock into loose material in a source area. The material is then transported from the source area to the deposition area. The type of sediment transported depends on the geology of the hinterland (the source area of the sediment). However, some sedimentary rocks, such as evaporites, are composed of material that form at the place of deposition. The nature of a sedimentary rock, therefore, not only depends on the sediment supply, but also on the sedimentary depositional environment in which it formed.

Cross-bedding and scour in a fine sandstone; the Logan Formation (Mississippian) of Jackson County, Ohio

Transformation (Diagenesis)

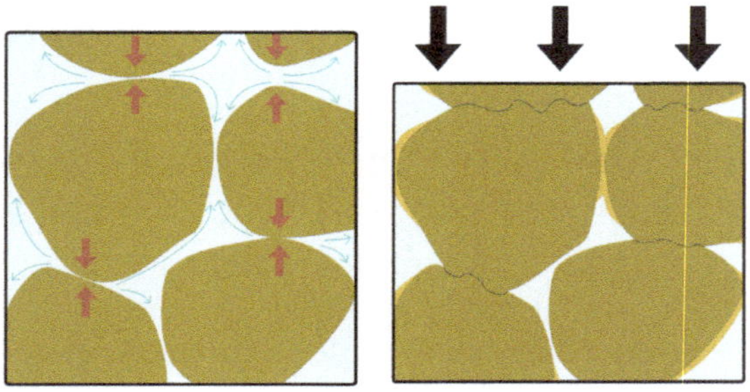

Pressure solution at work in a clastic rock. While material dissolves at places where grains are in contact, that material may recrystallize from the solution and act as cement in open pore spaces. As a result, there is a net flow of material from areas under high stress to those under low stress, producing a sedimentary rock becomes more compact and harder. Loose sand can become sandstone in this way.

The term diagenesis is used to describe all the chemical, physical, and biological changes, exclusive of surface weathering, undergone by a sediment after its initial deposition. Some of those processes cause the sediment to consolidate into a compact, solid substance from the originally loose material. Young sedimentary rocks, especially those of Quaternary age (the most recent period of the geologic time scale) are often still unconsolidated. As sediment deposition builds up, the overburden (lithostatic) pressure rises, and a process known as lithification takes place.

Sedimentary rocks are often saturated with seawater or groundwater, in which minerals can dissolve, or from which minerals can precipitate. Precipitating minerals reduce the pore space in a rock, a process called cementation. Due to the decrease in pore space, the original connate fluids are expelled. The precipitated minerals form a cement and make the rock more compact and competent. In this way, loose clasts in a sedimentary rock can become "glued" together.

When sedimentation continues, an older rock layer becomes buried deeper as a result. The lithostatic pressure in the rock increases due to the weight of the overlying sediment. This causes compaction, a process in which grains mechanically reorganize. Compaction is, for example, an important diagenetic process in clay, which can initially consist of 60% water. During compaction, this interstitial water is pressed out of pore spaces. Compaction can also be the result of dissolution of grains by pressure solution. The dissolved material precipitates again in open pore spaces, which means there is a net flow of material into the pores. However, in some cases, a certain mineral dissolves and does not precipitate again. This process, called leaching, increases pore space in the rock.

Some biochemical processes, like the activity of bacteria, can affect minerals in a rock and are therefore seen as part of diagenesis. Fungi and plants (by their roots) and various other organisms that live beneath the surface can also influence diagenesis.

Burial of rocks due to ongoing sedimentation leads to increased pressure and temperature, which stimulates certain chemical reactions. An example is the reactions by which organic material becomes lignite or coal. When temperature and pressure increase still further, the realm of diagenesis makes way for metamorphism, the process that forms metamorphic rock.

Properties

A piece of a banded iron formation, a type of rock that consists of alternating layers with iron(III) oxide (red) and iron(II) oxide (grey). BIFs were mostly formed during the Precambrian, when the atmosphere was not yet rich in oxygen. Moories Group, Barberton Greenstone Belt, South Africa

Color

The color of a sedimentary rock is often mostly determined by iron, an element with two major oxides: iron(II) oxide and iron(III) oxide. Iron(II) oxide (FeO) only forms under low oxygen (anoxic) circumstances and gives the rock a grey or greenish colour. Iron(III) oxide (Fe_2O_3) in a richer iron environment is often found in the form of the mineral hematite and gives the rock a reddish to brownish colour. In arid continental climates rocks are in direct contact with the atmosphere, and oxidation is an important process, giving the rock a red or orange colour. Thick sequences of red sedimentary rocks formed in arid climates are called red beds. However, a red colour does not necessarily mean the rock formed in a continental environment or arid climate.

The presence of organic material can colour a rock black or grey. Organic material is formed from dead organisms, mostly plants. Normally, such material eventually decays by oxidation or bacterial activity. Under anoxic circumstances, however, organic material cannot decay and leaves a dark sediment, rich in organic material. This can, for example, occur at the bottom of deep seas and lakes. There is little water mixing in such environments, as a result oxygen from surface water is not brought down, and the deposited sediment is normally a fine dark clay. Dark rocks, rich in organic material, are therefore often shales.

Texture

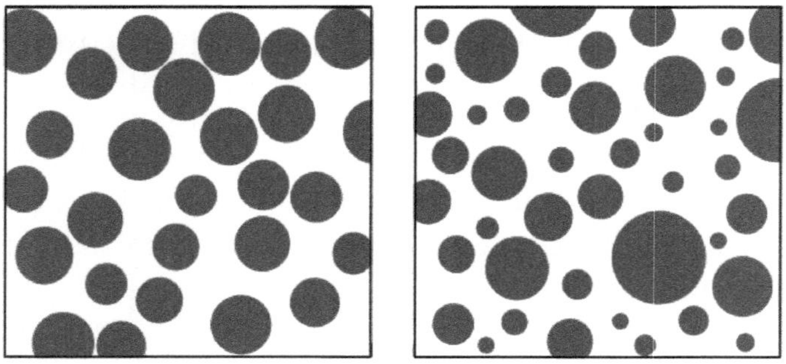

Diagram showing well-sorted (left) and poorly sorted (right) grains

The size, form and orientation of clasts (the original pieces of rock) in a sediment is called its texture. The texture is a small-scale property of a rock, but determines many of its large-scale properties, such as the density, porosity or permeability.

The 3D orientation of the clasts is called the fabric of the rock. Between the clasts, the rock can be composed of a matrix (a cement) that consists of crystals of one or more precipitated minerals. The size and form of clasts can be used to determine the velocity and direction of current in the sedimentary environment that moved the clasts from their origin; fine, calcareous mud only settles in quiet water while gravel and larger clasts are moved only by rapidly moving water. The grain size of a rock is usually expressed with the Wentworth scale, though alternative scales are sometimes used. The grain size can be expressed as a diameter or a volume, and is always an average value – a rock is composed of clasts with different sizes. The statistical distribution of grain sizes is different for different rock types and is described in a property called the sorting of the rock. When all clasts are more or less of the same size, the rock is called 'well-sorted', and when there is a large spread in grain size, the rock is called 'poorly sorted'.

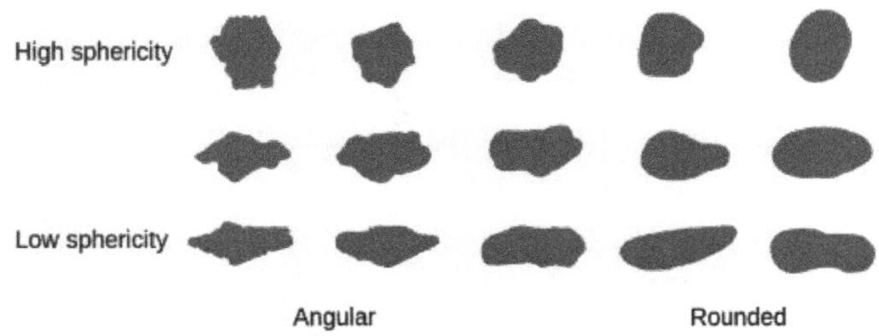

Diagram showing the rounding and sphericity of grains

The form of the clasts can reflect the origin of the rock.

Coquina, a rock composed of clasts of broken shells, can only form in energetic water. The form of a clast can be described by using four parameters:

- *Surface texture* describes the amount of small-scale relief of the surface of a grain that is too small to influence the general shape.
- *rounding* describes the general smoothness of the shape of a grain.
- 'Sphericity' describes the degree to which the grain approaches a sphere.
- 'Grain form' describes the three dimensional shape of the grain.

Chemical sedimentary rocks have a non-clastic texture, consisting entirely of crystals. To describe such a texture, only the average size of the crystals and the fabric are necessary.

Mineralogy

Most sedimentary rocks contain either quartz (especially siliciclastic rocks) or calcite (especially carbonate rocks). In contrast to igneous and metamorphic rocks, a sedimentary rock usually contains very few different major minerals. However, the origin of the minerals in a sedimentary rock is often more complex than in an igneous rock. Minerals in a sedimentary rock can have formed by precipitation during sedimentation or by diagenesis. In the second case, the mineral precipitate can have grown over an older generation of cement. A complex diagenetic history can be studied by optical mineralogy, using a petrographic microscope.

Carbonate rocks dominantly consist of carbonate minerals such as calcite, aragonite or dolomite. Both the cement and the clasts (including fossils and ooids) of a carbonate sedimentary rock can consist of carbonate minerals. The mineralogy of a clastic rock is determined by the material supplied by the source area, the manner of its transport to the place of deposition and the stability of that particular mineral. The resistance of rock forming minerals to weathering is expressed by Bowen's reaction series. In this series, quartz is the most stable, followed by feldspar, micas, and finally other less stable minerals that are only present when little weathering has occurred. The amount of weathering depends mainly on the distance to the source area, the local climate and the time it took for the sediment to be transported to the point where it is deposited. In most sedimentary rocks, mica, feldspar and less stable minerals have been reduced to clay minerals like kaolinite, illite or smectite.

Fossils

Fossil-rich layers in a sedimentary rock, Año Nuevo State Reserve, California

Among the three major types of rock, fossils are most commonly found in sedimentary rock. Unlike most igneous and metamorphic rocks, sedimentary rocks form at temperatures and pressures that do not destroy fossil remnants. Often these fossils may only be visible under magnification.

Dead organisms in nature are usually quickly removed by scavengers, bacteria, rotting and erosion, but sedimentation can contribute to exceptional circumstances where these natural processes are unable to work, causing fossilisation. The chance of fossilisation is higher when the sedimentation rate is high (so that a carcass is quickly buried), in anoxic environments (where little bacterial activity occurs) or when the organism had a particularly hard skeleton. Larger, well-preserved fossils are relatively rare.

Burrows in a turbidite, made by crustaceans, San Vincente Formation (early Eocene) of the Ainsa Basin, southern foreland of the Pyrenees

Fossils can be both the direct remains or imprints of organisms and their skeletons. Most commonly preserved are the harder parts of organisms such as bones, shells, and the woody tissue of plants. Soft tissue has a much smaller chance of being fossilized, and the preservation of soft tissue of animals older than 40 million years is very rare. Imprints of organisms made while they were still alive are called trace fossils, examples of which are burrows, footprints, etc.

As a part of a sedimentary or metamorphic rock, fossils undergo the same diagenetic processes as does the containing rock. A shell consisting of calcite can, for example, dissolve while a cement of silica then fills the cavity. In the same way, precipitating minerals can fill cavities formerly occupied by blood vessels, vascular tissue or other soft tissues. This preserves the form of the organism but changes the chemical composition, a process called permineralization. The most common minerals involved in permineralization are cements of carbonates (especially calcite), forms of amorphous silica (chalcedony, flint, chert) and pyrite. In the case of silica cements, the process is called lithification.

At high pressure and temperature, the organic material of a dead organism undergoes chemical reactions in which volatiles such as water and carbon dioxide are expulsed. The fossil, in the end, consists of a thin layer of pure carbon or its mineralized form, graphite. This form of fossilisation is called carbonisation. It is particularly important for plant fossils. The same process is responsible for the formation of fossil fuels like lignite or coal.

Primary Sedimentary Structures

Cross-bedding in a fluviatile sandstone, Middle Old Red Sandstone (Devonian) on Bressay, Shetland Islands

A flute cast, a type of sole marking, from the Book Cliffs of Utah

Ripple marks formed by a current in a sandstone that was later tilted (Haßberge, Bavaria)

Structures in sedimentary rocks can be divided into 'primary' structures (formed during deposition) and 'secondary' structures (formed after deposition). Unlike textures, structures are always large-scale features that can easily be studied in the field. Sedimentary structures can indicate something about the sedimentary environment or can serve to tell which side originally faced up where tectonics have tilted or overturned sedimentary layers.

Sedimentary rocks are laid down in layers called beds or strata. A bed is defined as a layer of rock that has a uniform lithology and texture. Beds form by the deposition of layers of sediment on top of each other. The sequence of beds that characterizes sedimentary rocks is called bedding. Single beds can be a couple of centimetres to several meters thick. Finer, less pronounced layers are called laminae, and the structure it forms in a rock is called lamination. Laminae are usually less than a few centimetres thick. Though bedding and lamination are often originally horizontal in nature, this is not always the case. In some environments, beds are deposited at a (usually small) angle. Sometimes multiple sets of layers with different orientations exist in the same rock, a structure called cross-bedding. Cross-bedding forms when small-scale erosion occurs during deposition, cutting off part of the beds. Newer beds then form at an angle to older ones.

The opposite of cross-bedding is parallel lamination, where all sedimentary layering is parallel. Differences in laminations are generally caused by cyclic changes in the sediment supply, caused, for example, by seasonal changes in rainfall, temperature or biochemical activity. Laminae that represent seasonal changes (similar to tree rings) are called varves. Any sedimentary rock composed of millimeter or finer scale layers can be named with the general term *laminite*. When sedimentary rocks have no lamination at all, their structural character is called massive bedding.

Graded bedding is a structure where beds with a smaller grain size occur on top of beds with larger grains. This structure forms when fast flowing water stops flowing. Larger, heavier clasts in suspension settle first, then smaller clasts. Although graded bedding can form in many different environments, it is a characteristic of turbidity currents.

The surface of a particular bed, called the bedform, can be indicative of a particular sedimentary environment, too. Examples of bed forms include dunes and ripple marks. Sole markings, such as tool marks and flute casts, are groves dug into a sedimentary layer that are preserved. These are often elongated structures and can be used to establish the direction of the flow during deposition.

Ripple marks also form in flowing water. There are two types of ripples: symmetric and asymmetric. Environments where the current is in one direction, such as rivers, produce asymmetric ripples. The longer flank of such ripples is on the upstream side of the current. Symmetric wave ripples occur in environments where currents reverse directions, such as tidal flats.

Mudcracks are a bed form caused by the dehydration of sediment that occasionally comes above the water surface. Such structures are commonly found at tidal flats or point bars along rivers.

Secondary Sedimentary Structures

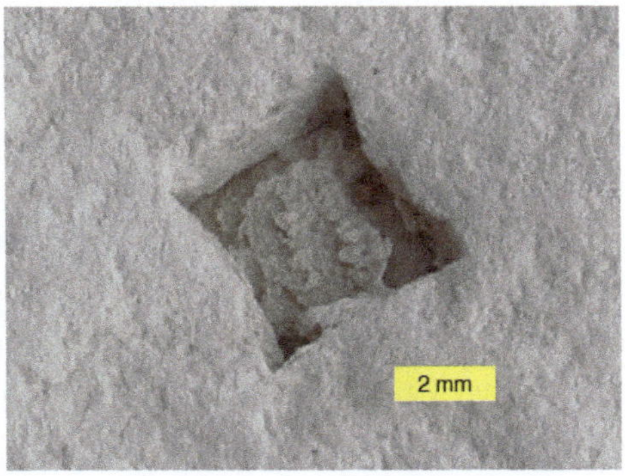

Halite crystal mold in dolomite, Paadla Formation (Silurian), Saaremaa, Estonia

Secondary sedimentary structures are those which formed after deposition. Such structures form by chemical, physical and biological processes within the sediment. They can be indicators of circumstances after deposition. Some can be used as way up criteria.

Organic materials in a sediment can leave more traces than just fossils. Preserved tracks and burrows are examples of trace fossils (also called ichnofossils). Such traces are relatively rare. Most trace fossils are burrows of molluscs or arthropods. This burrowing is called bioturbation by sedimentologists. It can be a valuable indicator of the biological and ecological environment that existed after the sediment was deposited. On the other hand, the burrowing activity of organisms can destroy other (primary) structures in the sediment, making a reconstruction more difficult.

Chert concretions in chalk, Middle Lefkara Formation (upper Paleocene to middle Eocene), Cyprus

Secondary structures can also form by diagenesis or the formation of a soil (pedogenesis) when a sediment is exposed above the water level. An example of a diagenetic structure common in carbonate rocks is a stylolite. Stylolites are irregular planes where material was dissolved into the pore fluids in the rock. This can result in the precipitation of a certain chemical species producing colouring and staining of the rock, or the formation of concretions. Concretions are roughly concentric bodies with a different composition from the host rock. Their formation can be the result of localized precipitation due to small differences in composition or porosity of the host rock, such as around fossils, inside burrows or around plant roots. In carbonate based rocks such as limestone or chalk, chert or flint concretions are common, while terrestrial sandstones can have iron concretions. Calcite concretions in clay are called septarian concretions.

After deposition, physical processes can deform the sediment, producing a third class of secondary structures. Density contrasts between different sedimentary layers, such as between sand and clay, can result in flame structures or load casts, formed by inverted diapirism. While the clastic bed is still fluid, diapirism can cause a denser upper layer to sink into a lower layer. Sometimes, density contrasts can result or grow when one of the lithologies dehydrates. Clay can be easily compressed as a result of dehydration, while sand retains the same volume and becomes relatively less dense. On the other hand, when the pore fluid pressure in a sand layer surpasses a critical point, the sand can break through overlying clay layers and flow through, forming discordant bodies of sedimentary rock called sedimentary dykes. The same process can form mud volcanoes on the surface where they broke through upper layers.

Sedimentary dykes can also be formed in a cold climate where the soil is permanently frozen during a large part of the year. Frost weathering can form cracks in the soil that fill with rubble from above. Such structures can be used as climate indicators as well as way up structures.

Density contrasts can also cause small-scale faulting, even while sedimentation progresses (synchronous-sedimentary faulting). Such faulting can also occur when large masses of non-lithified sediment are deposited on a slope, such as at the front side of a delta or the continental slope. Instabilities in such sediments can result in the deposited material to slump, producing fissures and folding. The resulting structures in the rock are syn-sedimentary folds and faults, which can be difficult to distinguish from folds and faults formed by tectonic forces acting on lithified rocks.

Sedimentary Environments

The setting in which a sedimentary rock forms is called the sedimentary environment. Every environment has a characteristic combination of geologic processes and circumstances. The type of sediment that is deposited is not only dependent on the sediment that is transported to a place, but also on the environment itself.

A marine environment means that the rock was formed in a sea or ocean. Often, a distinction is made between deep and shallow marine environments. Deep marine usually refers to environments more than 200 m below the water surface. Shallow marine environments exist adjacent to coastlines and can extend to the boundaries of the continental shelf. The water movements in such environments have a generally higher energy than that in deep environments, as wave activity diminishes with depth. This means that coarser sediment particles can be transported and the deposited sediment can be coarser than in deeper environments. When the sediment is transported from the continent, an alternation of sand, clay and silt is deposited. When the continent is far away, the amount of such sediment deposited may be small, and biochemical processes dominate the type of rock that forms. Especially in warm climates, shallow marine environments far offshore mainly see deposition of carbonate rocks. The shallow, warm water is an ideal habitat for many small organisms that build carbonate skeletons. When these organisms die, their skeletons sink to the bottom, forming a thick layer of calcareous mud that may lithify into limestone. Warm shallow marine environments also are ideal environments for coral reefs, where the sediment consists mainly of the calcareous skeletons of larger organisms.

In deep marine environments, the water current working the sea bottom is small. Only fine particles can be transported to such places. Typically sediments depositing on the ocean floor are fine clay or small skeletons of micro-organisms. At 4 km depth, the solubility of carbonates increases dramatically (the depth zone where this happens is called the lysocline). Calcareous sediment that sinks below the lysocline dissolves, as a result no limestone can be formed below this depth. Skeletons of micro-organisms formed of silica (such as radiolarians) are not as soluble and still deposit. An example of a rock formed of silica skeletons is radiolarite. When the bottom of the sea has a small inclination, for example at the continental slopes, the sedimentary cover can become unstable, causing turbidity currents. Turbidity currents are sudden disturbances of the normally quite deep marine environment and can cause the geologically speaking instantaneous deposition of large amounts of sediment, such as sand and silt. The rock sequence formed by a turbidity current is called a turbidite.

The coast is an environment dominated by wave action. At a beach, dominantly denser sediment such as sand or gravel, often mingled with shell fragments, is deposited, while the silt and clay sized material is kept in mechanical suspension. Tidal flats and shoals are places that sometimes dry because of the tide. They are often cross-cut by gullies, where the current is strong and the grain size of the deposited sediment is larger. Where rivers enter the body of water, either on a sea or lake coast, deltas can form. These are large accumulations of sediment transported from the continent to places in front of the mouth of the river. Deltas are dominantly composed of clastic sediment (in contrast to chemical).

A sedimentary rock formed on land has a continental sedimentary environment. Examples of continental environments are lagoons, lakes, swamps, floodplains and alluvial fans. In the quiet water of swamps, lakes and lagoons, fine sediment is deposited, mingled with organic material from dead plants and animals. In rivers, the energy of the water is much greater and can transport heavier clastic material. Besides transport by water, sediment can in continental environments also be transported by wind or glaciers. Sediment transported by wind is called aeolian and is always very well sorted, while sediment transported by a glacier is called glacial till and is characterized by very poor sorting.

Aeolian deposits can be quite striking. The depositional environment of the Touchet Formation, located in the Northwestern United States, had intervening periods of aridity which resulted in a series of rhythmite layers. Erosional cracks were later infilled with layers of soil material, especially from aeolian processes. The infilled sections formed vertical inclusions in the horizontally deposited layers of the Touchet Formation, and thus provided evidence of the events that intervened over time among the forty-one layers that were deposited.

Sedimentary Facies

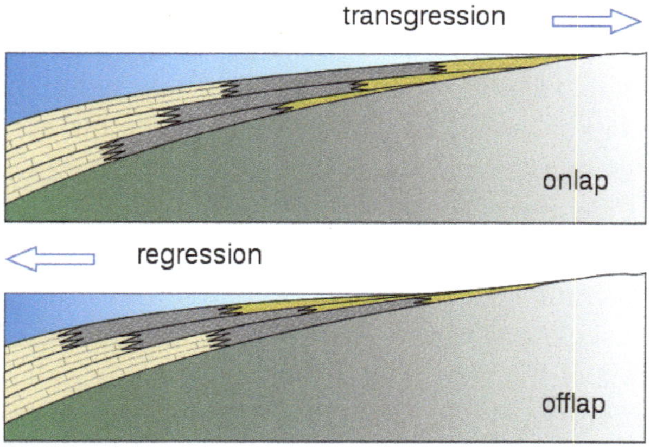

Shifting sedimentary facies in the case of transgression (above) and regression of the sea (below)

Sedimentary environments usually exist alongside each other in certain natural successions. A beach, where sand and gravel is deposited, is usually bounded by a deeper marine environment a little offshore, where finer sediments are deposited at the same time. Behind the beach, there can be dunes (where the dominant deposition is well sorted sand) or a lagoon (where fine clay and organic material is deposited). Every sedimentary environment has its own characteristic deposits. The typical rock formed in a certain environment is called its sedimentary facies. When sedimentary strata accumulate through time, the environment can shift, forming a change in facies in the subsurface at one location. On the other hand, when a rock layer with a certain age is followed laterally, the lithology (the type of rock) and facies eventually change.

Facies can be distinguished in a number of ways: the most common are by the lithology (for example: limestone, siltstone or sandstone) or by fossil content. Coral for example only lives in warm and shallow marine environments and fossils of coral are thus typical for shallow marine facies. Facies determined by lithology are called lithofacies; facies determined by fossils are biofacies.

Sedimentary environments can shift their geographical positions through time. Coastlines can shift in the direction of the sea when the sea level drops, when the surface rises due to tectonic forces in the Earth's crust or when a river forms a large delta. In the subsurface, such geographic shifts of sedimentary environments of the past are recorded in shifts in sedimentary facies. This means that sedimentary facies can change either parallel or perpendicular to an imaginary layer of rock with a fixed age, a phenomenon described by Walther's Law.

The situation in which coastlines move in the direction of the continent is called transgression. In the case of transgression, deeper marine facies are deposited over shallower facies, a succession called onlap. Regression is the situation in which a coastline moves in the direction of the sea. With regression, shallower facies are deposited on top of deeper facies, a situation called offlap.

The facies of all rocks of a certain age can be plotted on a map to give an overview of the palaeogeography. A sequence of maps for different ages can give an insight in the development of the regional geography.

Sedimentary Basins

Places where large-scale sedimentation takes place are called sedimentary basins. The amount of sediment that can be deposited in a basin depends on the depth of the basin, the so-called accommodation space. The depth, shape and size of a basin depend on tectonics, movements within the Earth's lithosphere. Where the lithosphere moves upward (tectonic uplift), land eventually rises above sea level, so that and erosion removes material, and the area becomes a source for new sediment. Where the lithosphere moves downward (tectonic subsidence), a basin forms and sedimentation can take place. When the lithosphere keeps subsiding, new accommodation space keeps being created.

A type of basin formed by the moving apart of two pieces of a continent is called a rift basin. Rift basins are elongated, narrow and deep basins. Due to divergent movement, the lithosphere is stretched and thinned, so that the hot asthenosphere rises and heats the overlying rift basin. Apart from continental sediments, rift basins normally also have part of their infill consisting of volcanic deposits. When the basin grows due to continued stretching of the lithosphere, the rift grows and the sea can enter, forming marine deposits.

When a piece of lithosphere that was heated and stretched cools again, its density rises, causing isostatic subsidence. If this subsidence continues long enough, the basin is called a sag basin. Examples of sag basins are the regions along passive continental margins, but sag basins can also be found in the interior of continents. In sag basins, the extra weight of the newly deposited sediments is enough to keep the subsidence going in a vicious circle. The total thickness of the sedimentary infill in a sag basins can thus exceed 10 km.

A third type of basin exists along convergent plate boundaries - places where one tectonic plate moves under another into the asthenosphere. The subducting plate bends and forms a fore-arc basin in front of the overriding plate—an elongated, deep asymmetric basin. Fore-arc basins are filled with deep marine deposits and thick sequences of turbidites. Such infill is called flysch. When the convergent movement of the two plates results in continental collision, the basin becomes shallower and develops into a foreland basin. At the same time, tectonic uplift forms a mountain belt in the overriding plate, from which large amounts of material are eroded and transported to the basin. Such erosional material of a growing mountain chain is called molasse and has either a shallow marine or a continental facies.

At the same time, the growing weight of the mountain belt can cause isostatic subsidence in the area of the overriding plate on the other side to the mountain belt. The basin type resulting from this subsidence is called a back-arc basin and is usually filled by shallow marine deposits and molasse.

Cyclic alternation of competent and less competent beds in the Blue Lias at Lyme Regis, southern England

Influence of Astronomical Cycles

In many cases facies changes and other lithological features in sequences of sedimentary rock have a cyclic nature. This cyclic nature was caused by cyclic changes in sediment supply and the sedimentary environment. Most of these cyclic changes are caused by astronomic cycles. Short astronomic cycles can be the difference between the tides or the spring tide every two weeks. On a larger time-scale, cyclic changes in climate and sea level are caused by Milankovitch cycles: cyclic changes in the orientation and/or position of the Earth's rotational axis and orbit around the Sun. There are a number of Milankovitch cycles known, lasting between 10,000 and 200,000 years.

Relatively small changes in the orientation of the Earth's axis or length of the seasons can be a major influence on the Earth's climate. An example are the ice ages of the past 2.6 million years (the Quaternary period), which are assumed to have been caused by astronomic cycles. Climate change can influence the global sea level (and thus the amount of accommodation space in sedimentary basins) and sediment supply from a certain region. Eventually, small changes in astronomic parameters can cause large changes in sedimentary environment and sedimentation.

Sedimentation Rates

The rate at which sediment is deposited differs depending on the location. A channel in a tidal flat can see the deposition of a few metres of sediment in one day, while on the deep ocean floor each year only a few millimetres of sediment accumulate. A distinction can be made between normal sedimentation and sedimentation caused by catastrophic processes. The latter category includes all kinds of sudden exceptional processes like mass movements, rock slides or flooding. Catastrophic processes can see the sudden deposition of a large amount of sediment at once. In some sedimentary environments, most of the total column of sedimentary rock was formed by catastrophic processes, even though the environment is usually a quiet place. Other sedimentary environments are dominated by normal, ongoing sedimentation.

In many cases, sedimentation occurs slowly. In a desert, for example, the wind deposits siliciclastic material (sand or silt) in some spots, or catastrophic flooding of a wadi may cause sudden deposits of large quantities of detrital material, but in most places eolian erosion dominates. The amount of sedimentary rock that forms is not only dependent on the amount of supplied material, but also on how well the material consolidates. Erosion removes most deposited sediment shortly after deposition.

Stratigraphy

That new rock layers are above older rock layers is stated in the principle of superposition. There are usually some gaps in the sequence called unconformities. These represent periods where no new sediments were laid down, or when earlier sedimentary layers were raised above sea level and eroded away.

The Permian through Jurassic stratigraphy of the Colorado Plateau area of southeastern Utah that makes up much of the famous prominent rock formations in protected areas such as Capitol Reef National Park and Canyonlands National Park. From top to bottom: Rounded tan domes of the Navajo Sandstone, layered red Kayenta Formation, cliff-forming, vertically jointed, red Wingate Sandstone, slope-forming, purplish Chinle Formation, layered, lighter-red Moenkopi Formation, and white, layered Cutler Formation sandstone. Picture from Glen Canyon National Recreation Area, Utah.

Sedimentary rocks contain important information about the history of the Earth. They contain fossils, the preserved remains of ancient plants and animals. Coal is considered a type of sedimentary rock. The composition of sediments provides us with clues as to the original rock. Differences between successive layers indicate changes to the environment over time. Sedimentary rocks can contain fossils because, unlike most igneous and metamorphic rocks, they form at temperatures and pressures that do not destroy fossil remains.

Metamorphic Rock

Quartzite, a form of metamorphic rock, from the Museum of Geology at University of Tartu collection.

Metamorphic rocks arise from the transformation of existing rock types, in a process called metamorphism, which means "change in form". The original rock (protolith) is subjected to heat (temperatures greater than 150 to 200 °C) and pressure (1500 bars), causing profound physical and/or chemical change. The protolith may be a sedimentary rock, an igneous rock or another older metamorphic rock.

Metamorphic rocks make up a large part of the Earth's crust and are classified by texture and by chemical and mineral assemblage (metamorphic facies). They may be formed simply by being deep beneath the Earth's surface, subjected to high temperatures and the great pressure of the rock layers above it. They can form from tectonic processes such as continental collisions, which cause horizontal pressure, friction and distortion. They are also formed when rock is heated up by the intrusion of hot molten rock called magma from the Earth's interior. The study of metamorphic rocks (now exposed at the Earth's surface following erosion and uplift) provides information about the temperatures and pressures that occur at great depths within the Earth's crust. Some examples of metamorphic rocks are gneiss, slate, marble, schist, and quartzite.

Metamorphic Minerals

Metamorphic minerals are those that form only at the high temperatures and pressures associated with the process of metamorphism. These minerals, known as index minerals, include sillimanite, kyanite, staurolite, andalusite, and some garnet.

Other minerals, such as olivines, pyroxenes, amphiboles, micas, feldspars, and quartz, may be found in metamorphic rocks, but are not necessarily the result of the process of metamorphism. These minerals formed during the crystallization of igneous rocks. They are stable at high temperatures and pressures and may remain chemically unchanged during the metamorphic process. However, all minerals are stable only within certain limits, and the presence of some minerals in metamorphic rocks indicates the approximate temperatures and pressures at which they formed.

The change in the particle size of the rock during the process of metamorphism is called recrystallization. For instance, the small calcite crystals in the sedimentary rock limestone and chalk change into larger crystals in the metamorphic rock marble, or in metamorphosed sandstone, recrystallization of the original quartz sand grains results in very compact quartzite, also known as metaquartzite, in which the often larger quartz crystals are interlocked. Both high temperatures and pressures contribute to recrystallization. High temperatures allow the atoms and ions in solid crystals to migrate, thus reorganizing the crystals, while high pressures cause solution of the crystals within the rock at their point of contact.

Foliation

The layering within metamorphic rocks is called *foliation* (derived from the Latin word *folia*, meaning "leaves"), and it occurs when a rock is being shortened along one axis during recrystallization. This causes the platy or elongated crystals of minerals, such as mica and chlorite, to become rotated such that their long axes are perpendicular to the orientation of shortening. This results in a banded, or foliated rock, with the bands showing the colors of the minerals that formed them.

Folded foliation in a metamorphic rock from near Geirangerfjord, Norway

Textures are separated into foliated and non-foliated categories. Foliated rock is a product of differential stress that deforms the rock in one plane, sometimes creating a plane of cleavage. For example, slate is a foliated metamorphic rock, originating from shale. Non-foliated rock does not have planar patterns of strain.

Rocks that were subjected to uniform pressure from all sides, or those that lack minerals with distinctive growth habits, will not be foliated. Where a rock has been subject to differential stress, the type of foliation that develops depends on the metamorphic grade. For instance, starting with a mudstone, the following sequence develops with increasing temperature: slate is a very fine-grained, foliated metamorphic rock, characteristic of very low grade metamorphism, while phyllite is fine-grained and found in areas of low grade metamorphism, schist is medium to coarse-grained and found in areas of medium grade metamorphism, and gneiss coarse to very coarse-grained, found in areas of high-grade metamorphism. Marble is generally not foliated, which allows its use as a material for sculpture and architecture.

Another important mechanism of metamorphism is that of chemical reactions that occur between minerals without them melting. In the process atoms are exchanged between the minerals, and thus new minerals are formed. Many complex high-temperature reactions may take place, and each mineral assemblage produced provides us with a clue as to the temperatures and pressures at the time of metamorphism.

Metasomatism is the drastic change in the bulk chemical composition of a rock that often occurs during the processes of metamorphism. It is due to the introduction of chemicals from other surrounding rocks. Water may transport these chemicals rapidly over great distances. Because of the role played by water, metamorphic rocks generally contain many elements absent from the original rock, and lack some that originally were present. Still, the introduction of new chemicals is not necessary for recrystallization to occur.

Types of Metamorphism

Contact Metamorphism

A contact metamorphic rock made of interlayered calcite and serpentine from the Precambrian of Canada. Once thought to be a fossil called *Eozoön canadense*. Scale in mm.

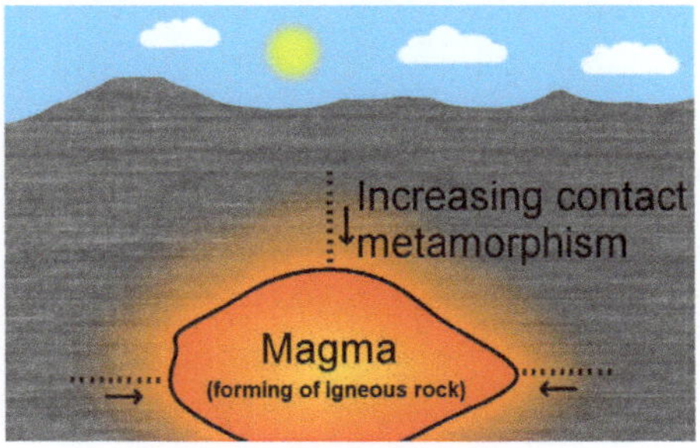

Contact metamorphism is the name given to the changes that take place when magma is injected into the surrounding solid rock (country rock). The changes that occur are greatest wherever the magma comes into contact with the rock because the temperatures are highest at this boundary and decrease with distance from it. Around the igneous rock that forms from the cooling magma is a metamorphosed zone called a *contact metamorphism aureole*. Aureoles may show all degrees of metamorphism from the contact area to unmetamorphosed (unchanged) country rock some distance away. The formation of important ore minerals may occur by the process of metasomatism at or near the contact zone.

When a rock is contact altered by an igneous intrusion it very frequently becomes more indurated, and more coarsely crystalline. Many altered rocks of this type were formerly called hornstones, and the term *hornfels* is often used by geologists to signify those fine grained, compact, non-foliated products of contact metamorphism. A shale may become a dark argillaceous hornfels, full of tiny plates of brownish biotite; a marl or impure limestone may change to a grey, yellow or greenish lime-silicate-hornfels or siliceous marble, tough and splintery, with abundant augite, garnet, wollastonite and other minerals in which calcite is an important component. A diabase or andesite may become a diabase hornfels or andesite hornfels with development of new hornblende and biotite and a partial recrystallization of the original feldspar. Chert or flint may become a finely crystalline quartz rock; sandstones lose their clastic structure and are converted into a mosaic of small close-fitting grains of quartz in a metamorphic rock called quartzite.

If the rock was originally banded or foliated (as, for example, a laminated sandstone or a foliated calc-schist) this character may not be obliterated, and a banded hornfels is the product; fossils even may have their shapes preserved, though entirely recrystallized, and in many contact-altered lavas the vesicles are still visible, though their contents have usually entered into new combinations to form minerals that were not originally present. The minute structures, however, disappear, often completely, if the thermal alteration is very profound. Thus small grains of quartz in a shale are lost or blend with the surrounding particles of clay, and the fine ground-mass of lavas is entirely reconstructed.

By recrystallization in this manner peculiar rocks of very distinct types are often produced. Thus shales may pass into cordierite rocks, or may show large crystals of andalusite (and chiastolite), staurolite, garnet, kyanite and sillimanite, all derived from the aluminous content of the original shale. A considerable amount of mica (both muscovite and biotite) is often simultaneously formed, and the resulting product has a close resemblance to many kinds of schist. Limestones, if pure, are often turned into coarsely crystalline marbles; but if there was an admixture of clay or sand in the original rock such minerals as garnet, epidote, idocrase, wollastonite, will be present. Sandstones when greatly heated may change into coarse quartzites composed of large clear grains of quartz. These more intense stages of alteration are not so commonly seen in igneous rocks, because their minerals, being formed at high temperatures, are not so easily transformed or recrystallized.

In a few cases rocks are fused and in the dark glassy product minute crystals of spinel, sillimanite and cordierite may separate out. Shales are occasionally thus altered by basalt dikes, and feldspathic sandstones may be completely vitrified. Similar changes may be induced in shales by the burning of coal seams or even by an ordinary furnace.

There is also a tendency for metasomatism between the igneous magma and sedimentary country rock, whereby the chemicals in each are exchanged or introduced into the other. Granites may absorb fragments of shale or pieces of basalt. In that case, hybrid rocks called skarn arise, which don't have the characteristics of normal igneous or sedimentary rocks. Sometimes an invading granite magma permeates the rocks around, filling their joints and planes of bedding, etc., with threads of quartz and feldspar. This is very exceptional but instances of it are known and it may take place on a large scale.

Regional Metamorphism

Regional metamorphism, also known as dynamic metamorphism, is the name given to changes in great masses of rock over a wide area. Rocks can be metamorphosed simply by being at great depths below the Earth's surface, subjected to high temperatures and the great pressure caused by the immense weight of the rock layers above. Much of the lower continental crust is metamorphic, except for recent igneous intrusions. Horizontal tectonic movements such as the collision of continents create orogenic belts, and cause high temperatures, pressures and deformation in the rocks along these belts. If the metamorphosed rocks are later uplifted and exposed by erosion, they may occur in long belts or other large areas at the surface. The process of metamorphism may have destroyed the original features that could have revealed the rock's previous history. Recrystallization of the rock will destroy the textures and fossils present in sedimentary rocks. Metasomatism will change the original composition.

Mississippian marble in Big Cottonwood Canyon, Wasatch Mountains, Utah.

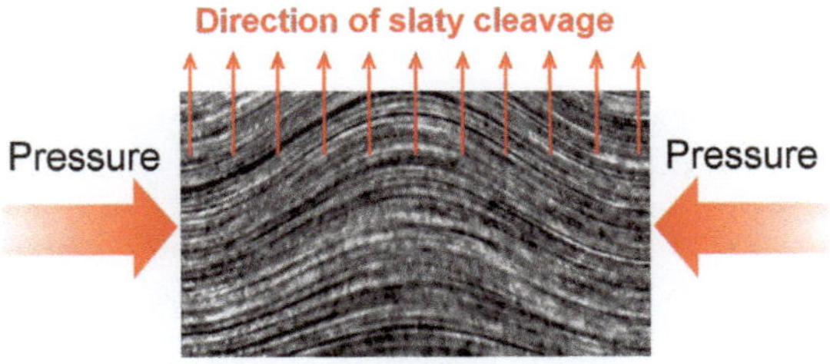

Dynamic metamorphism

Regional metamorphism tends to make the rock more indurated and at the same time to give it a foliated, shistose or gneissic texture, consisting of a planar arrangement of the minerals, so that platy or prismatic minerals like mica and hornblende have their longest axes arranged parallel to one another. For that reason many of these rocks split readily in one direction along mica-bearing zones (schists). In gneisses, minerals also tend to be segregated into bands; thus there are seams of quartz and of mica in a mica schist, very thin, but consisting essentially of one mineral. Along the mineral layers composed of soft or fissile minerals the rocks will split most readily, and the freshly split specimens will appear to be faced or coated with this mineral; for example, a piece of mica schist looked at facewise might be supposed to consist entirely of shining scales of mica. On the edge of the specimens, however, the white folia of granular quartz will be visible. In gneisses these alternating folia are sometimes thicker and less regular than in schists, but most importantly less micaceous; they may be lenticular, dying out rapidly. Gneisses also, as a rule, contain more feldspar than schists do, and are tougher and less fissile. Contortion or crumbling of the foliation is by no means uncommon; splitting faces are undulose or puckered. Schistosity and gneissic banding (the two main types of foliation) are formed by directed pressure at elevated temperature, and to interstitial movement, or internal flow arranging the mineral particles while they are crystallizing in that directed pressure field.

Rocks that were originally sedimentary and rocks that were undoubtedly igneous may be metamorphosed into schists and gneisses. If originally of similar composition they may be very difficult to distinguish from one another if the metamorphism has been great. A quartz-porphyry, for example, and a fine feldspathic sandstone, may both be metamorphosed into a grey or pink mica-schist.

Metamorphic Rock Textures

The five basic metamorphic textures with typical rock types are slaty (includes slate and phyllite; the foliation is called "slaty cleavage"), schistose (includes schist; the foliation is called "schistosity"), gneissose (gneiss; the foliation is called "gneissosity"), granoblastic (includes granulite, some marbles and quartzite), and hornfelsic (includes hornfels and skarn).

References

- Prothero, Donald R.; Schwab, Fred (2004). Sedimentary geology : an introduction to sedimentary rocks and stratigraphy (2nd ed.). New York: Freeman. p. 12. ISBN 978-0-7167-3905-0.
- Geoff C. Brown; C. J. Hawkesworth; R. C. L. Wilson (1992). Understanding the Earth (2nd ed.). Cambridge University Press. p. 93. ISBN 0-521-42740-1.
- Foulger, G.R. (2010). Plates vs. Plumes: A Geological Controversy. Wiley-Blackwell. ISBN 978-1-4051-6148-0.
- Wicander R. & Munroe J. (2005). Essentials of Geology. Cengage Learning. pp. 174–177. ISBN 9780495013655.

6

Various Volcanic Eruptions

Broadly there are three types of volcanic eruptions, namely, magmatic eruptions, phreato-magmatic eruptions and phreatic eruptions. These broadly defined types have several sub-types such as vulcanian eruptions, plinian eruptions and surtseyan eruptions. Volcanology is best understood in confluence with the major topics listed in the following chapter.

Types of Volcanic Eruptions

Several types of volcanic eruptions—during which lava, tephra (ash, lapilli, volcanic bombs and blocks), and assorted gases are expelled from a volcanic vent or fissure—have been distinguished by volcanologists. These are often named after famous volcanoes where that type of behavior has been observed. Some volcanoes may exhibit only one characteristic type of eruption during a period of activity, while others may display an entire sequence of types all in one eruptive series.

Some of the eruptive structures formed during volcanic activity: a Plinian eruption column, Hawaiian pahoehoe flows, and a lava arc from a Strombolian eruption.

There are three different types of eruptions. The most well-observed are magmatic eruptions, which involve the decompression of gas within magma that propels it forward. Phreatomagmatic eruptions are another type of volcanic eruption, driven by the compression of gas within magma, the direct opposite of the process powering magmatic activity. The third eruptive type is the phreatic eruption, which is driven by the superheating of steam via contact with magma; these eruptive types often exhibit no magmatic release, instead causing the granulation of existing rock.

Within these wide-defining eruptive types are several subtypes. The weakest are Hawaiian and submarine, then Strombolian, followed by Vulcanian and Surtseyan. The stronger eruptive types are Pelean eruptions, followed by Plinian eruptions; the strongest eruptions are called "Ultra-Plinian." Subglacial and phreatic eruptions are defined by their eruptive mechanism, and vary in strength. An important measure of eruptive strength is Volcanic Explosivity Index (VEI), an order of magnitude scale ranging from 0 to 8 that often correlates to eruptive types.

Eruption Mechanisms

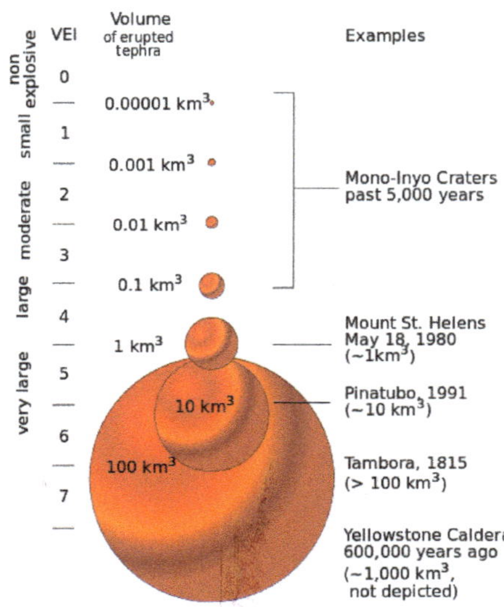

Diagram showing the scale of VEI correlation with total ejecta volume.

Volcanic eruptions arise through three main mechanisms:

- Gas release under decompression causing magmatic eruptions
- Thermal contraction from chilling on contact with water causing phreatomagmatic eruptions
- Ejection of entrained particles during steam eruptions causing phreatic eruptions

There are two types of eruptions in terms of activity, explosive eruptions and effusive eruptions. Explosive eruptions are characterized by gas-driven explosions that propels magma and tephra. Effusive eruptions, meanwhile, are characterized by the outpouring of lava without significant explosive eruption.

Volcanic eruptions vary widely in strength. On the one extreme there are effusive Hawaiian eruptions, which are characterized by lava fountains and fluid lava flows, which are typically not very dangerous. On the other extreme, Plinian eruptions are large, violent, and highly dangerous explosive events. Volcanoes are not bound to one eruptive style, and frequently display many different types, both passive and explosive, even the span of a single eruptive cycle. Volcanoes do not always erupt vertically from a single crater near their peak, either. Some volcanoes exhibit lateral and fissure eruptions. Notably, many Hawaiian eruptions start from rift zones, and some of the strongest Surtseyan eruptions develop along fracture zones. Scientists believed that pulses of magma mixed together in the chamber before climbing upward—a process estimated to take several thousands of years. But Columbia University volcanologists found that the eruption of Costa Rica's Irazú Volcano in 1963 was likely triggered by magma that took a nonstop route from the mantle over just a few months.

Volcano Explosivity Index

The volcanic explosivity index (commonly shortened to VEI) is a scale, from 0 to 8, for measuring the strength of eruptions. It is used by the Smithsonian Institution's Global Volcanism Program in assessing the impact of historic and prehistoric lava flows. It operates in a way similar to the Richter scale for earthquakes, in that each interval in value represents a tenfold increasing in magnitude (it is logarithmic). The vast majority of volcanic eruptions are of VEIs between 0 and 2.

Magmatic Eruptions

Magmatic eruptions produce juvenile clasts during explosive decompression from gas release. They range in intensity from the relatively small lava fountains on Hawaii to catastrophic Ultra-Plinian eruption columns more than 30 km (19 mi) high, bigger than the eruption of Mount Vesuvius in 79 that buried Pompeii.

Hawaiian

Hawaiian eruptions are a type of volcanic eruption, named after the Hawaiian volcanoes with which this eruptive type is hallmark. Hawaiian eruptions are the calmest types of volcanic events, characterized by the effusive eruption of very fluid basalt-type lavas with low gaseous content. The volume of ejected material from Hawaiian eruptions is less than half of that found in other eruptive types. Steady production of small amounts of lava builds up the large, broad form of a shield volcano. Eruptions are not centralized at the main summit as with other volcanic types, and often occur at vents around the summit and from fissure vents radiating out of the center.

Hawaiian eruptions often begin as a line of vent eruptions along a fissure vent, a so-called "curtain of fire." These die down as the lava begins to concentrate at a few of the vents. Central-vent eruptions, meanwhile, often take the form of large lava fountains (both continuous and sporadic), which can reach heights of hundreds of meters or more. The particles from lava fountains usually cool in the air before hitting the ground, resulting in the accumulation of cindery scoria fragments; however, when the air is especially thick with clasts, they cannot cool off fast enough due to the surrounding heat, and hit the ground still hot, the accumulation of which forms spatter cones. If eruptive rates are high enough, they may even form splatter-fed lava flows. Hawaiian eruptions are often extremely long lived; Pu'u'Ō'ō, a cinder cone of Kilauea, has been erupting continuously since 1983. Another Hawaiian volcanic feature is the formation of active lava lakes, self-maintaining pools of raw lava with a thin crust of semi-cooled rock; there are currently only 5 such lakes in the world, and the one at Kīlauea's Kupaianaha vent is one of them.

Flows from Hawaiian eruptions are basaltic, and can be divided into two types by their structural characteristics. Pahoehoe lava is a relatively smooth lava flow that can be billowy or ropey. They can move as one sheet, by the advancement of "toes," or as a snaking lava column. A'a lava flows are denser and more viscous then pahoehoe, and tend to move slower. Flows can measure 2 to 20 m (7 to 66 ft) thick. A'a flows are so thick that the outside layers cools into a rubble-like mass, insulating the still-hot interior and preventing it from cooling. A'a lava moves in a peculiar way—the front of the flow steepens due to pressure from behind until it breaks off, after which the general mass behind it moves forward. Pahoehoe lava can sometimes become A'a lava due to increasing viscosity or increasing rate of shear, but A'a lava never turns into pahoehoe flow.

Hawaiian eruptions are responsible for several unique volcanological objects. Small volcanic particles are carried and formed by the wind, chilling quickly into teardrop-shaped glassy fragments known as Pele's tears (after Pele, the Hawaiian volcano deity). During especially high winds these chunks may even take the form of long drawn-out strands, known as Pele's hair. Sometimes basalt aerates into reticulite, the lowest density rock type on earth.

Although Hawaiian eruptions are named after the volcanoes of Hawaii, they are not necessarily restricted to them; the largest lava fountain ever recorded formed on the island of Izu Ōshima (on Mount Mihara) in 1986, a 1,600 m (5,249 ft) gusher that was more than twice as high as the mountain itself (which stands at 764 m (2,507 ft)).

Volcanoes known to have Hawaiian activity include:

- Pu'u'Ō'ō, a parasitic cinder cone located on Kilauea on the island of Hawaiʻi which has been erupting continuously since 1983. The eruptions began with a 6 km (4 mi)-long fissure-based "curtain of fire" on 3 January. These gave way to centralized eruptions on the site of Kilauea's east rift, eventually building up the still active cone.
- For a list of all of the volcanoes of Hawaii, List of volcanoes in the Hawaiian - Emperor seamount chain.

- Mount Etna, Italy.
- Mount Mihara in 1986

Strombolian

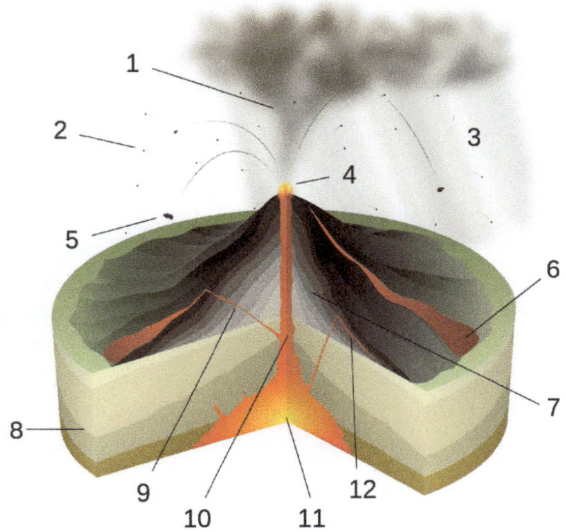

Diagram of a Strombolian eruption. (key: 1. Ash plume 2. Lapilli 3. Volcanic ash rain 4. Lava fountain 5. Volcanic bomb 6. Lava flow 7. Layers of lava and ash 8. Stratum 9. Dike 10. Magma conduit 11. Magma chamber 12. Sill) Click for larger version.

Strombolian eruptions are a type of volcanic eruption, named after the volcano Stromboli, which has been erupting continuously for centuries. Strombolian eruptions are driven by the bursting of gas bubbles within the magma. These gas bubbles within the magma accumulate and coalesce into large bubbles, called gas slugs. These grow large enough to rise through the lava column. Upon reaching the surface, the difference in air pressure causes the bubble to burst with a loud pop, throwing magma in the air in a way similar to a soap bubble. Because of the high gas pressures associated with the lavas, continued activity is generally in the form of episodic explosive eruptions accompanied by the distinctive loud blasts. During eruptions, these blasts occur as often as every few minutes.

The term "Strombolian" has been used indiscriminately to describe a wide variety of volcanic eruptions, varying from small volcanic blasts to large eruptive columns. In reality, true Strombolian eruptions are characterized by short-lived and explosive eruptions of lavas with intermediate viscosity, often ejected high into the air. Columns can measure hundreds of meters in height. The lavas formed by Strombolian eruptions are a form of relatively viscous basaltic lava, and its end product is mostly scoria. The relative passivity of Strombolian eruptions, and its non-damaging nature to its source vent allow Strombolian eruptions to continue unabated for thousands of years, and also makes it one of the least dangerous eruptive types.

An example of the lava arcs formed during Strombolian activity. This image is of Stromboli itself.

Strombolian eruptions eject volcanic bombs and lapilli fragments that travel in parabolic paths before landing around their source vent. The steady accumulation of small fragments builds cinder cones composed completely of basaltic pyroclasts. This form of accumulation tends to result in well-ordered rings of tephra.

Strombolian eruptions are similar to Hawaiian eruptions, but there are differences. Strombolian eruptions are noisier, produce no sustained eruptive columns, do not produce some volcanic products associated with Hawaiian volcanism (specifically Pele's tears and Pele's hair), and produce fewer molten lava flows (although the eruptive material does tend to form small rivulets).

Volcanoes known to have Strombolian activity include:

- Parícutin, Mexico, which erupted from a fissure in a cornfield in 1943. Two years into its life, pyroclastic activity began to wane, and the outpouring of lava from its base became its primary mode of activity. Eruptions ceased in 1952, and the final height was 424 m (1,391 ft). This was the first time that scientists are able to observe the complete life cycle of a volcano.
- Mount Etna, Italy, which has displayed Strombolian activity in recent eruptions, for example in 1981, 1999, 2002-2003, and 2009.
- Mount Erebus in Antarctica, the southernmost active volcano in the world, having been observed erupting since 1972. Eruptive activity at Erebus consists of frequent Strombolian activity.
- Stromboli itself. The namesake of the mild explosive activity that it possesses has been active throughout historical time; essentially continuous Strombolian eruptions, occasionally accompanied by lava flows, have been recorded at Stromboli for more than a millennium.

Vulcanian

Vulcanian eruptions are a type of volcanic eruption, named after the volcano Vulcano. It was named so following Giuseppe Mercalli's observations of its 1888-1890 eruptions. In Vulcanian eruptions, highly viscous magma within the volcano make it difficult for vesiculate gases to escape. Similar to Strombolian eruptions, this leads to the buildup of high gas pressure, eventually popping the cap holding the magma down and resulting in an explosive eruption. However, unlike Strombolian eruptions, ejected lava fragments are not aerodynamic; this is due to the higher viscosity of Vulcanian magma and the greater incorporation of crystalline material broken off from the former cap. They are also more explosive than their Strombolian counterparts, with eruptive columns often reaching between 5 and 10 km (3 and 6 mi) high. Lastly, Vulcanian deposits are andesitic to dacitic rather than basaltic.

Initial Vulcanian activity is characterized by a series of short-lived explosions, lasting a few minutes to a few hours and typified by the ejection of volcanic bombs and blocks. These eruptions wear down the lava dome holding the magma down, and it disintegrates, leading to much more quiet and continuous eruptions. Thus an early sign of future Vulcanian activity is lava dome growth, and its collapse generates an outpouring of pyroclastic material down the volcano's slope.

Tavurvur in Papua New Guinea erupting.

Deposits near the source vent consist of large volcanic blocks and bombs, with so-called "bread-crust bombs" being especially common. These deeply cracked volcanic chunks form when the exterior of ejected lava cools quickly into a glassy or fine-grained shell, but the inside continues to cool and vesiculate. The center of the fragment expands, cracking the exterior. However the bulk of Vulcanian deposits are fine grained ash. The ash is only moderately dispersed, and its abundance indicates a high degree of fragmentation, the result of high gas contents within the magma. In some cases these have been found to be the result of interaction with meteoric water, suggesting that Vulcanian eruptions are partially hydrovolcanic.

Volcanoes that have exhibited Vulcanian activity include:

- Sakurajima, Japan has been the site of Vulcanian activity near-continuously since 1955.

- Tavurvur, Papua New Guinea, one of several volcanoes in the Rabaul Caldera.
- Irazú Volcano in Costa Rica exhibited Vulcanian activity in its 1965 eruption.

Peléan

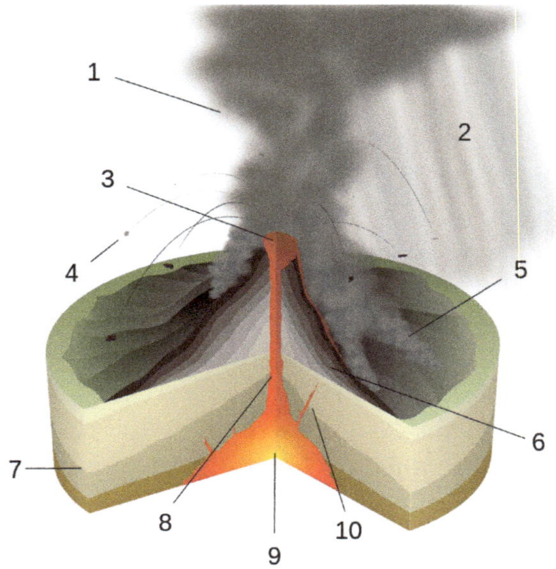

Diagram of Peléan eruption. (key: 1. Ash plume 2. Volcanic ash rain 3. Lava dome 4. Volcanic bomb 5. Pyroclastic flow 6. Layers of lava and ash 7. Stratum 8. Magma conduit 9. Magma chamber 10. Dike) Click for larger version.

Peléan eruptions (or nuée ardente) are a type of volcanic eruption, named after the volcano Mount Pelée in Martinique, the site of a massive Peléan eruption in 1902 that is one of the worst natural disasters in history. In Peléan eruptions, a large amount of gas, dust, ash, and lava fragments are blown out the volcano's central crater, driven by the collapse of rhyolite, dacite, and andesite lava dome collapses that often create large eruptive columns. An early sign of a coming eruption is the growth of a so-called Peléan or lava spine, a bulge in the volcano's summit preempting its total collapse. The material collapses upon itself, forming a fast-moving pyroclastic flow (known as a block-and-ash flow) that moves down the side of the mountain at tremendous speeds, often over 150 km (93 mi) per hour. These massive landslides make Peléan eruptions one of the most dangerous in the world, capable of tearing through populated areas and causing massive loss of life. The 1902 eruption of Mount Pelée caused tremendous destruction, killing more than 30,000 people and competely destroying the town of St. Pierre, the worst volcanic event in the 20th century.

Peléan eruptions are characterized most prominently by the incandescent pyroclastic flows that they drive. The mechanics of a Peléan eruption are very similar to that of a Vulcanian eruption, except that in Peléan eruptions the volcano's structure is able to withstand more pressure, hence the eruption occurs as one large explosion rather than several smaller ones.

Volcanoes known to have Peléan activity include:

- Mount Pelée, Martinique. The 1902 eruption of Mount Pelée completely devastated the island, destroying the town of St. Pierre and leaving only 3 survivors. The eruption was directly preceded by lava dome growth.
- Mayon Volcano, the Philippines most active volcano. It has been the site of many different types of eruptions, Peléan included. Approximately 40 ravines radiate from the summit and provide pathways for frequent pyroclastic flows and mudslides to the lowlands below. Mayon's most violent eruption occurred in 1814 and was responsible for over 1200 deaths.
- The 1951 Peléan eruption of Mount Lamington. Prior to this eruption the peak had not even been recognized as a volcano. Over 3,000 people were killed, and it has become a benchmark for studying large Peléan eruptions.

Pyroclastic flows at Mayon Volcano, Philippines, 1984.

The lava spine that developed after the 1902 eruption of Mount Pelée

Mount Lamington following the devastating 1951 eruption.

Plinian

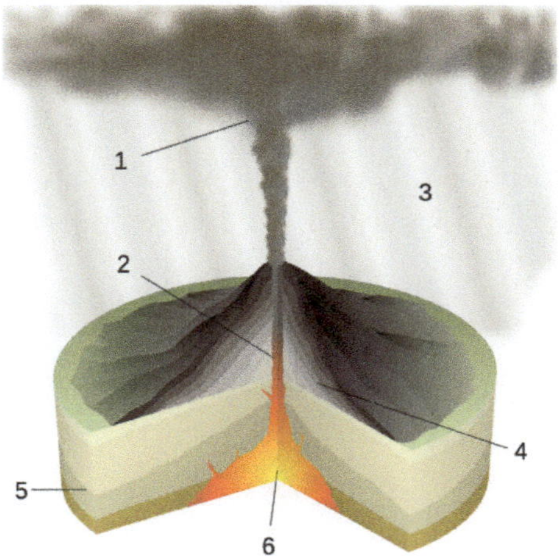

Diagram of a Plinian eruption. (key: 1. Ash plume 2. Magma conduit 3. Volcanic ash rain 4. Layers of lava and ash 5. Stratum 6. Magma chamber) Click for larger version.

Plinian eruptions (or Vesuvian) are a type of volcanic eruption, named for the historical eruption of Mount Vesuvius in 79 of Mount Vesuvius that buried the Roman towns of Pompeii and Herculaneum and, specifically, for its chronicler Pliny the Younger. The process powering Plinian eruptions starts in the magma chamber, where dissolved volatile gases are stored in the magma. The gases vesiculate and accumulate as they rise through the magma conduit. These bubbles agglutinate and once they reach a certain size (about 75% of the total volume of the magma conduit) they explode. The narrow confines of the conduit force the gases and associated magma up, forming an eruptive column. Eruption velocity is controlled by the gas contents of the column, and low-strength surface rocks commonly crack under the pressure of the eruption, forming a flared outgoing structure that pushes the gases even faster.

These massive eruptive columns are the distinctive feature of a Plinian eruption, and reach up 2 to 45 km (1 to 28 mi) into the atmosphere. The densest part of the plume, directly above the volcano, is driven internally by gas expansion. As it reaches higher into the air the plume expands and becomes less dense, convection and thermal expansion of volcanic ash drive it even further up into the stratosphere. At the top of the plume, powerful prevailing winds drive the plume in a direction away from the volcano.

These highly explosive eruptions are associated with volatile-rich dacitic to rhyolitic lavas, and occur most typically at stratovolcanoes. Eruptions can last anywhere from hours to days, with longer eruptions being associated with more felsic volcanoes. Although they are associated with felsic magma, Plinian eruptions can just as well occur at basaltic volcanoes, given that the magma chamber differentiates and has a structure rich in silicon dioxide.

Plinian eruptions are similar to both Vulcanian and Strombolian eruptions, except that rather than creating discrete explosive events, Plinian eruptions form sustained eruptive columns. They are also similar to Hawaiian lava fountains in that both eruptive types produce sustained eruption columns maintained by the growth of bubbles that move up at about the same speed as the magma surrounding them.

Regions affected by Plinian eruptions are subjected to heavy pumice airfall affecting an area 0.5 to 50 km^3 (0 to 12 cu mi) in size. The material in the ash plume eventually finds its way back to the ground, covering the landscape in a thick layer of many cubic kilometers of ash.

Lahar flows from the 1985 eruption of Nevado del Ruiz, which totally destroyed the town of Armero in Colombia.

However the most dangerous eruptive feature are the pyroclastic flows generated by material collapse, which move down the side of the mountain at extreme speeds of up to 700 km (435 mi) per hour and with the ability to extend the reach of the eruption hundreds of kilometers. The ejection of hot material from the volcano's summit melts snowbanks and ice deposits on the volcano, which mixes with tephra to form lahars, fast moving mudslides with the consistency of wet concrete that move at the speed of a river rapid.

Major Plinian eruptive events include:

- The AD 79 eruption of Mount Vesuvius buried the Roman towns of Pompeii and Herculaneum under a layer of ash and tephra. It is the model Plinian eruption. Mount Vesuvius has erupted several times since then. Its last eruption was in 1944 and caused problems for the allied armies as they advanced through Italy. It was the report by Pliny that Younger that lead scientists to refer to vesuvian eruptions as "Plinian".
- The 1980 eruption of Mount St. Helens in Washington, which ripped apart the volcano's summit, was a Plinian eruption of Volcanic Explosivity Index (**VEI**) 5.
- The strongest types of eruptions, with a VEI of 8, are so-called "Ultra-Plinian" eruptions, such as the most recent one at Lake Toba 74 thousand years ago, which put out 2800 times the material erupted by Mount St. Helens in 1980.
- Hekla in Iceland, an example of basaltic Plinian volcanism being its 1947-48 eruption. The past 800 years have been a pattern of violent initial eruptions of pumice followed by prolonged extrusion of basaltic lava from the lower part of the volcano.
- Pinatubo in the Philippines on 15 June 1991, which produced 5 km^3 (1 cu mi) of dacitic magma, a 40 km (25 mi) high eruption column, and released 17 megatons of sulfur dioxide.

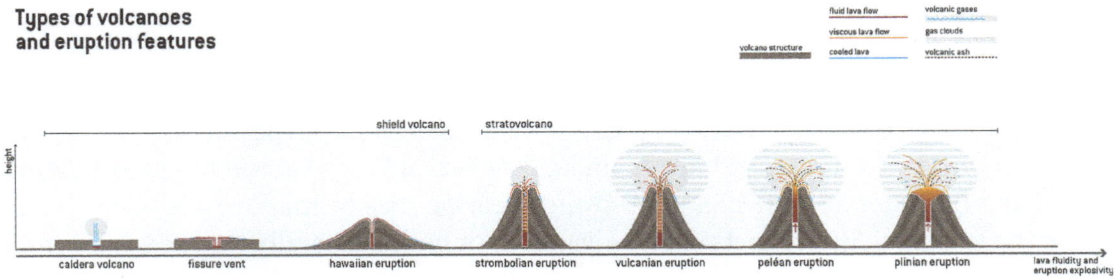

Phreatomagmatic Eruptions

Phreatomagmatic eruptions are eruptions that arise from interactions between water and magma. They are driven from thermal contraction (as opposed to magmatic eruptions, which are driven by thermal expansion) of magma when it comes in contact with water. This temperature difference between the two causes violent water-lava interactions that make up the eruption. The products of phreatomagmatic eruptions are believed to be more regular in shape and finer grained than the products of magmatic eruptions because of the differences in eruptive mechanisms.

There is debate about the exact nature of phreatomagmatic eruptions, and some scientists believe that fuel-coolant reactions may be more critical to the explosive nature than thermal contraction. Fuel coolant reactions may fragment the volcanic material by propagating stress waves, widening cracks and increasing surface area that ultimetly lead to rapid cooling and explosive contraction-driven eruptions.

Surtseyan

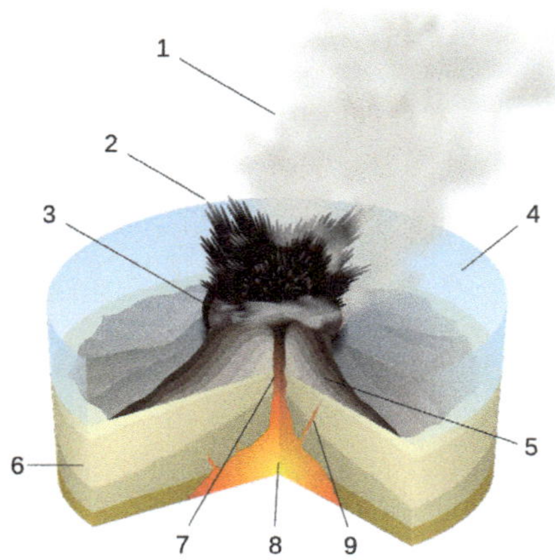

Diagram of a Surtseyan eruption. (key: 1. Water vapor cloud 2. Compressed ash 3. Crater 4. Water 5. Layers of lava and ash 6. Stratum 7. Magma conduit 8. Magma chamber 9. Dike)

A Surtseyan eruption (or hydrovolcanic) is a type of volcanic eruption caused by shallow-water interactions between water and lava, named so after its most famous example, the eruption and formation of the island of Surtsey off the coast of Iceland in 1963. Surtseyan eruptions are the "wet" equivalent of ground-based Strombolian eruptions, but because of where they are taking place they are much more explosive. This is because as water is heated by lava, it flashes in steam and expands violently, fragmenting the magma it is in contact with into fine-grained ash. Surtseyan eruptions are the hallmark of shallow-water volcanic oceanic islands, however they are not specifically confined to them. Surtseyan eruptions can happen on land as well, and are caused by rising magma that comes into contact with an aquifer (water-bearing rock formation) at shallow levels under the volcano. The products of Surtseyan eruptions are generally oxidized palagonite basalts (though andesitic eruptions do occur, albeit rarely), and like Strombolian eruptions Surtseyan eruptions are generally continuous or otherwise rhythmic.

A distinct defining feature of a Surtseyan eruption is the formation of a pyroclastic surge (or *base surge*), a ground hugging radial cloud that develops along with the eruption column. Base surges are caused by the gravitational collapse of a vaporous eruptive column, one that is denser overall then a regular volcanic column. The densest part of the cloud is nearest to the vent, resulting a wedge shape. Associated with these laterally moving rings are dune-shaped depositions of rock left behind by the lateral movement. These are occasionally disrupted by bomb sags, rock that was flung out by the explosive eruption and followed a ballistic path to the ground. Accumulations of wet, spherical ash known as accretionary lapilli is another common surge indicator.

Over time Surtseyan eruptions tend to form maars, broad low-relief volcanic craters dug into the ground, and tuff rings, circular structures built of rapidly quenched lava. These structures are associated with a single vent eruption, however if eruptions arise along fracture zones a rift zone may be dug out; these eruptions tend to be more violent then the ones forming a tuff ring or maars, an example being the 1886 eruption of Mount Tarawera. Littoral cones are another hydrovolcanic feature, generated by the explosive deposition of basaltic tephra (although they are not truly volcanic vents). They form when lava accumulates within cracks in lava, superheats and explodes in a steam explosion, breaking the rock apart and depositing it on the volcano's flank. Consecutive explosions of this type eventually generate the cone.

Volcanoes known to have Surtseyan activity include:

- Surtsey, Iceland. The volcano built itself up from depth and emerged above the Atlantic Ocean off the coast of Iceland in 1963. Initial hydrovolcanics were highly explosive, but as the volcano grew out rising lava started to interact less with the water and more with the air, until finally Surtseyan activity waned and became more Strombolian in character.
- Ukinrek Maars in Alaska, 1977, and Capelinhos in the Azores, 1957, both examples of above-water Surtseyan activity.
- Mount Tarawera in New Zealand erupted along a rift zone in 1886, killing 150 people.
- Ferdinandea, a seamount in the Mediterranean Sea, breached sea level in July 1831 and was the source of a dispute over sovereignty between Italy, France, and Great Britain. The volcano did not build tuff cones strongly enough to withstand erosion, and disappeared back below the waves soon after it appeared.
- The underwater volcano Hunga Tonga in Tonga breached sea level in 2009. Both of its vents exhibited Surtseyan activity for much of the time. It was also the site of an earlier eruption in May 1988.

Surtsey, erupting 13 days after breaching the water. A tuff ring surrounds the vent.

The fissure formed by the 1886 eruption of Mount Tarawera, an example of a fracture zone eruption.

Submarine

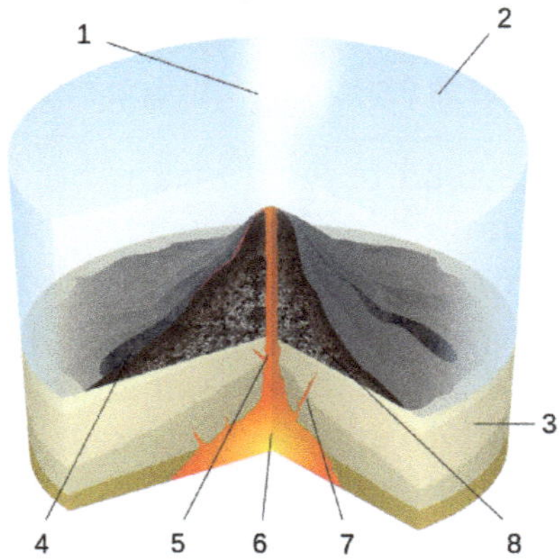

Diagram of a Submarine eruption. (key: 1. Water vapor cloud 2. Water 3. Stratum 4. Lava flow 5. Magma conduit 6. Magma chamber 7. Dike 8. Pillow lava) Click to enlarge.

Submarine eruptions are a type of volcanic eruption that occurs underwater. An estimated 75% of the total volcanic eruptive volume is generated by submarine eruptions near mid ocean ridges alone, however because of the problems associated with detecting deep sea volcanics, they remained virtually unknown until advances in the 1990s made it possible to observe them.

Submarine eruptions may produce seamounts which may break the surface to form volcanic islands and island chains.

Submarine volcanism is driven by various processes. Volcanoes near plate boundaries and mid-ocean ridges are built by the decompression melting of mantle rock that rises on an upwelling portion of a convection cell to the crustal surface. Eruptions associated with subducting zones, meanwhile, are driven by subducting plates that add volatiles to the rising plate, lowering its melting point. Each process generates different rock; mid-ocean ridge volcanics are primarily basaltic, whereas subduction flows are mostly calc-alkaline, and more explosive and viscous.

Spreading rates along mid-ocean ridges vary widely, from 2 cm (0.8 in) per year at the Mid-Atlantic Ridge, to up to 16 cm (6 in) along the East Pacific Rise. Higher spreading rates are a probably cause for higher levels of volcanism. The technology for studying seamount eruptions did not exist until advancements in hydrophone technology made it possible to "listen" to acoustic waves, known as T-waves, released by submarine earthquakes associated with submarine volcanic eruptions. The reason for this is that land-based seismometers cannot detect sea-based earthquakes below a magnitude of 4, but acoustic waves travel well in water and long periods of time. A system in the North Pacific, maintained by the United States Navy and originally intended for the detection of submarines, has detected an event on average every 2 to 3 years.

The most common underwater flow is pillow lava, a circular lava flow named after its unusual shape. Less common are glassy, marginal sheet flows, indicative of larger-scale flows. Volcaniclastic sedimentary rocks are common in shallow-water environments. As plate movement starts to carry the volcanoes away from their eruptive source, eruption rates start to die down, and water erosion grinds the volcano down. The final stages of eruption caps the seamount in alkalic flows. There are about 100,000 deepwater volcanoes in the world, although most are beyond the active stage of their life. Some exemplary seamounts are Loihi Seamount, Bowie Seamount, Davidson Seamount, and Axial Seamount.

Subglacial

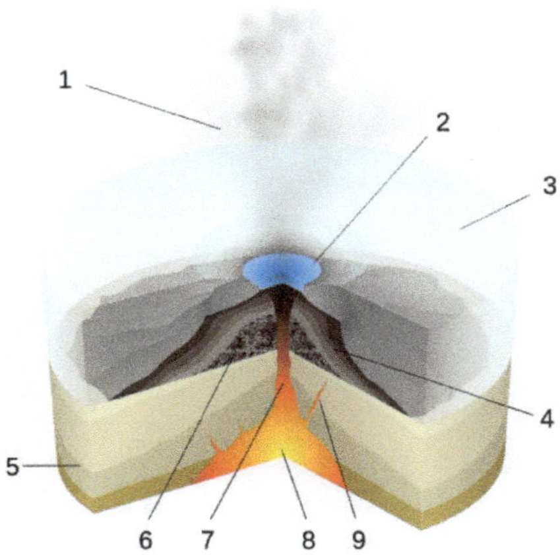

A diagram of a Subglacial eruption. (key: 1. Water vapor cloud 2. Crater lake 3. Ice 4. Layers of lava and ash 5. Stratum 6. Pillow lava 7. Magma conduit 8. Magma chamber 9. Dike) Click for larger version.

Subglacial eruptions are a type of volcanic eruption characterized by interactions between lava and ice, often under a glacier. The nature of glaciovolcanism dictates that it occurs at areas of high latitude and high altitude. It has been suggested that subglacial volcanoes that are not actively erupting often dump heat into the ice covering them, producing meltwater. This meltwater mix means that subglacial eruptions often generate dangerous jökulhlaups (floods) and lahars.

The study of glaciovolcanism is still a relatively new field. Early accounts described the unusual flat-topped steep-sided volcanoes (called tuyas) in Iceland that were suggested to have formed from eruptions below ice. The first English-language paper on the subject was published in 1947 by William Henry Mathews, describing the Tuya Butte field in northwest British Columbia, Canada. The eruptive process that builds these structures, originally inferred in the paper, begins with volcanic growth below the glacier. At first the eruptions resemble those that occur in the deep sea, forming piles of pillow lava at the base of the volcanic structure. Some of the lava shatters when it comes in contact with the cold ice, forming a glassy breccia called hyaloclastite. After a while the ice finally melts into a lake, and the more explosive eruptions of Surtseyan activity begins, building up flanks made up of mostly hyaloclastite. Eventually the lake boils off from continued volcanism, and the lava flows become more effusive and thicken as the lava cools much more slowly, often forming columnar jointing. Well-preserved tuyas show all of these stages, for example Hjorleifshofdi in Iceland.

Products of volcano-ice interactions stand as various structures, whose shape is dependent on complex eruptive and environmental interactions. Glacial volcanism is a good indicator of past ice distribution, making it an important climatic marker. Since they are imbedded in ice, as ice retracts worldwide there are concerns that tuyas and other structures may destabalize, resulting in mass landslides. Evidence of volcanic-glacial interactions are evident in Iceland and parts of British Columbia, and it is even possible that they play a role in deglaciation.

Herðubreið, a tuya in Iceland.

Glaciovolcanic products have been identified in Iceland, the Canadian province of British Columbia, the U.S. states of Hawaii and Alaska, the Cascade Range of western North America, South America and even on the planet Mars. Volcanoes known to have subglacial activity include:

- Mauna Kea in tropical Hawaii. There is evidence of past subglacial eruptive activity on the volcano in the form of a subglacial deposit on its summit. The eruptions originated about 10,000 years ago, during the last ice age, when the summit of Mauna Kea was covered in ice.
- In 2008, the British Antarctic Survey reported a volcanic eruption under the Antarctica ice sheet 2,200 years ago. It is believed to be that this was the biggest eruption in Antarctica in the last 10,000 years. Volcanic ash deposits from the volcano were identified through an airborne radar survey, buried under later snowfalls in the Hudson Mountains, close to Pine Island Glacier.
- Iceland, well known for both glaciers and volcanoes, is often a site of subglacial eruptions. An example an eruption under the Vatnajökull ice cap in 1996, which occurred under an estimated 2,500 ft (762 m) of ice.
- As part of the search for life on Mars, scientists have suggested that there may be subglacial volcanoes on the red planet. Several potential sites of such volca-

nism have been reviewed, and compared extensively with similar features in Iceland:

Viable microbial communities have been found living in deep (-2800 m) geothermal groundwater at 349 K and pressures >300 bar. Furthermore, microbes have been postulated to exist in basaltic rocks in rinds of altered volcanic glass. All of these conditions could exist in polar regions of Mars today where subglacial volcanism has occurred.

Phreatic Eruptions

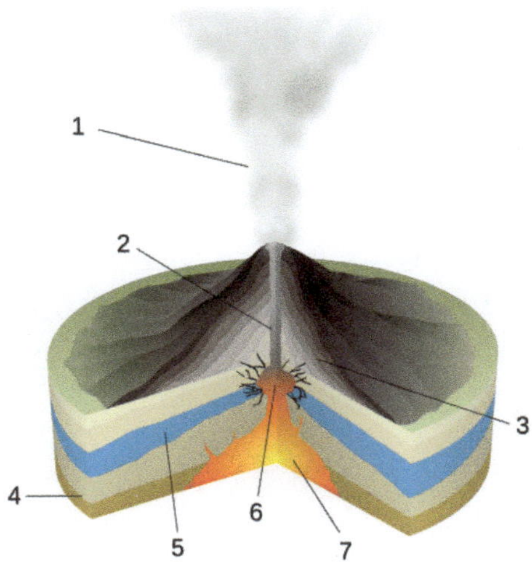

Diagram of a phreatic eruption. (key: 1. Water vapor cloud 2. Magma conduit 3. Layers of lava and ash 4. Stratum 5. Water table 6. Explosion 7. Magma chamber)

Phreatic eruptions (or steam-blast eruptions) are a type of eruption driven by the expansion of steam. When cold ground or surface water come into contact with hot rock or magma it superheats and explodes, fracturing the surrounding rock and thrusting out a mixture of steam, water, ash, volcanic bombs, and volcanic blocks. The distinguishing feature of phreatic explosions is that they only blast out fragments of pre-existing solid rock from the volcanic conduit; no new magma is erupted. Because they are driven by the cracking of rock strata under pressure, phreatic activity does not always result in an eruption; if the rock face is strong enough to withstand the explosive force, outright eruptions may not occur, although cracks in the rock will probably develop and weaken it, furthering future eruptions.

Often a precursor of future volcanic activity, phreatic eruptions are generally weak, although there have been exceptions. Some phreatic events may be triggered by earthquake activity, another volcanic precursor, and they may also travel along dike lines. Phreatic eruptions form base surges, lahars, avalanches, and volcanic block “rain.” They may also release deadly toxic gas able to suffocate anyone in range of the eruption.

Volcanoes known to exhibit phreatic activity include:

- Mount St. Helens, which exhibited phreatic activity just prior to its catastrophic 1980 eruption (which was itself Plinian).
- Taal Volcano, Philippines, 1965.
- La Soufrière of Guadeloupe (Lesser Antilles), 1975-1976 activity.
- Soufrière Hills volcano on Montserrat, West Indies, 1995–2012.
- Poás Volcano, has frequent geyser like phreatic eruptions from its crater lake.
- Mount Bulusan, well known for its sudden phreatic eruptions.
- Mount Ontake, all historical eruptions of this volcano have been phreatic including the deadly 2014 eruption.

Explosive Eruption

Mount Saint Helens explosive eruption on July 22, 1980

An explosive eruption is a volcanic term to describe a violent, explosive type of eruption. Mount St. Helens in 1980 was an example. Such eruptions result when sufficient gas has dissolved under pressure within a viscous magma such that expelled lava violently froths into volcanic ash when pressure is suddenly lowered at the vent. Sometimes a lava plug will block the conduit to the summit, and when this occurs, eruptions are more violent. Explosive eruptions can send rocks, dust, gas and pyroclastic material up to 20 km into the atmosphere at rate of up to 100,000 tonnes per second, traveling at several hundred meters per second. This cloud will then collapse, creating a pyroclástic flow of hot volcanic matter.

Stages of an Explosive Eruption

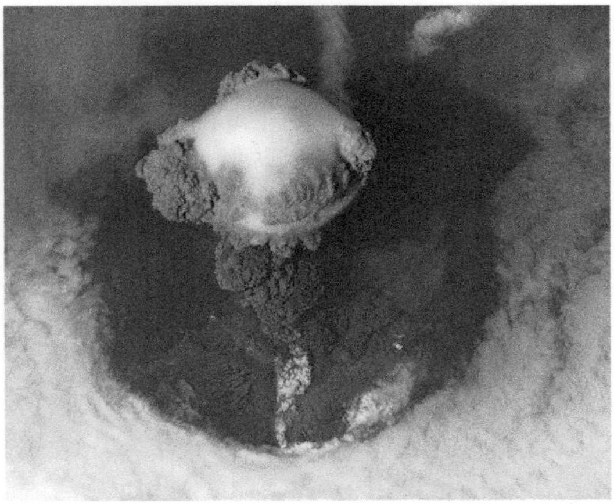

An early stage of the July 12, 2009 eruption of Sarychev volcano, seen from space

An explosive eruption always begins with some form of blockage in the crater of a volcano that prevents the release of gases trapped in highly viscous andesitic or rhyolitic magma. The high viscosity of these forms of magma prevents the release of trapped gases. When this type of magma flows towards the surface pressure builds, eventually causing the blockage to be blasted out in an explosive eruption. The pressure from the magma and gases are released through the weakest point in the cone, usually the crater. However, in the case of the eruption of Mount St. Helens, pressure was released through the side of the volcano, rather than the crater.

The sudden release of pressure causes the gases in the magma to suddenly froth and create volcanic ash and pumice, which is then ejected through the volcanic vent to create the signature eruption column commonly associated with explosive eruptions. The size and duration of the column depends on the volume of magma being released and how much pressure the magma was under.

Types Of Explosive Eruption

1. Vulcanian eruption
2. Peléan eruption
3. Plinian eruption
4. Phreatomagmatic eruption and Phreatic eruption
    1. Surtseyan eruption

- Consequences:
    - Eruption column
    - Pyroclastic flow

- Pyroclastic fall
- Pyroclastic surge

Pyroclastic Flows

Pyroclastic flows occur towards the end of an explosive eruption, as pressure begins to decline. The eruption column of ash is supported by pressure from the gases being released, and as the gases are depleted, pressure falls and the eruption column begins to collapse. When the column collapses in on itself, ash and rock fall back down to the ground and begin to flow down the slopes of the volcano. These flows can travel at up to 80 km per hour, and reach temperatures of 200° to 700° Celsius. The high temperatures can cause combustion of any flammable materials in its path, including wood, vegetation, and buildings. When snow and ice melt as a part of an eruption, large amounts of water mixed in with the flow can create lahars. The risk of lahars is particularly high on volcanoes such as Mount Rainier near Seattle and Tacoma, Washington.

Supervolcanoes

The eruption of supervolcanoes is the rarest of volcanic eruptions but also the most destructive. The timescale between these eruptions is generally marked by hundreds or thousands of years. This type of eruption generally causes destruction on a continental scale, and can also result in the lowering of temperatures worldwide.

Phreatomagmatic Eruption

Ashfall deposit of phreatomagmatic origin, overlying magmatic lapilli fall deposit of magmatic origin.

Phreatomagmatic eruptions are volcanic eruptions resulting from interaction between magma and water. They differ from exclusively magmatic eruptions and phreatic eruptions. Unlike phreatic eruptions, the products of phreatomagmatic eruptions contain juvenile (magmatic) clasts. It is common for a large explosive eruption to have magmatic and phreatomagmatic components.

Mechanisms

Several competing theories exist as to the exact mechanism of ash formation. The most common is the theory of explosive thermal contraction of particles under rapid cooling from contact with water. In many cases the water is supplied by the sea, for example with Surtsey. In other cases the water may be present in a lake or caldera-lake, for example Santorini, where the phreatomagmatic component of the Minoan eruption was a result of both a lake and later the sea. There have also been examples of interaction between magma and water in an aquifer. Many of the cinder cones on Tenerife are believed to be phreatomagmatic because of these circumstances.

The other competing theory is based on fuel-coolant reactions, which have been modeled for the nuclear industry. Under this theory the fuel (in this case, the magma) fragments upon contact with a coolant (the sea, a lake or aquifer). The propagating stress waves and thermal contraction widen cracks and increase the interaction surface area, leading to explosively rapid cooling rates. The two mechanisms proposed are very similar and the reality is most likely a combination of both.

Deposits

Phreatomagmatic ash is formed by the same mechanisms across a wide range of compositions, basic and acidic. Blocky and equant clasts with low vesicule content are formed. The deposits of phreatomagmatic explosive eruptions are also believed to be better sorted and finer grained than the deposits of magmatic eruption. This is a result of the much higher fragmentation of phreatomagmatic eruptions.

Hyaloclastite

Hyaloclastite is glass found with pillow basalts that were produced by non-explosive quenching and fracturing of basaltic glass. These are still classed as phreatomagmatic eruptions, as they produce juvenile clasts from the interaction of water and magma. They can be formed at water depths of >500 m, where hydrostatic pressure is high enough to inhibit vesiculation in basaltic magma.

Hyalotuff

Hyalotuff is a type of rock formed by the explosive fragmentation of glass during phreatomagmatic eruptions at shallow water depths (or within aquifers). Hyalotuffs have a layered nature that is believed to be a result of dampened oscillation in discharge rate, with a period of several minutes. The deposits are much finer grained than the deposits of magmatic eruptions, due to the much higher fragmentation of the type of eruption. The deposits appear better sorted than magmatic deposits in the field because of their fine nature, but grain size analysis reveals that the deposits are much more poorly sorted than their magmatic counterparts. A clast known as an accretionary lapilli is distinctive to phreatomagmatic deposits, and is a major factor for identification in the field. Accretionary lapilli form as a result of the cohesive properties of wet ash, causing the particles to bind. They have a circular structure when specimens are viewed in hand and under the microscope.

A further control on the morphology and characteristics of a deposit is the water to magma ratio. It is believed that the products of phreatomagmatic eruptions are fine grained and poorly sorted where the magma/water ratio is high, but when there is a lower magma/water ratio the deposits may be coarser and better sorted.

Surface Features

There are two types of vent landforms from the explosive interaction of magma and ground or surface water; *tuff cones* and *tuff rings*. Both of the landforms are associated with monogenetic volcanoes and polygenetic volcanoes. In the case of polygenetic volcanoes they are often interbedded with lavas, ignimbrites and ash- and lapilli-fall deposits. It is expected that tuff rings and tuff cones might be present on the surface of Mars.

Crest of old tuff ring, including part of the maar crater of a monogenetic volcano, Tenerife, Canary Islands. The maar crater has been used for agriculture.

Tuff Rings

Koko Crater is an old extinct tuff cone in the Hawaiian Island of Oahu.

Tuff rings have a low profile apron of tephra surrounding a wide crater (called a *maar* crater) that is generally lower than the surrounding topography. The tephra is often unaltered and thinly bedded, and is generally considered to be an ignimbrite, or the product of a pyroclastic density current. They are built around a volcanic vent located in a lake, coastal zone, marsh or an area of abundant groundwater.

Tuff Cones

Tuff cones are steep sloped and cone shaped. They have wide craters and are formed of highly altered, thickly bedded tephra. They are considered to be a taller variant of a tuff ring, formed by less powerful eruptions. Tuff cones are usually small in height. Koko Crater is 1,208 feet.

Examples of Phreatomagmatic Eruptions

Fort Rock, an eroded tuff ring in Oregon, USA.

Minoan Eruption of Santorini

Santorini is part of the Southern Aegean volcanic arc, 140 km north of Crete. The Minoan eruption of Santorini, was the latest eruption and occurred in the first half of the 17th century BC. The eruption was of predominantly rhyodacite composition. The Minoan eruption had four phases. Phase 1 was a white to pink pumice fallout with dispersal axis trending ESE. The deposit has a maximum thickness of 6 m and ash flow layers are interbedded at the top. Phase 2 has ash and lapilli beds that are cross stratified with mega-ripples and dune like structures. The deposit thicknesses vary from 10 cm to 12 m. Phases 3 and 4 are pyroclastic density current deposits. Phases 1 and 3 were phreatomagmatic.

Pinatubo, 1991

Mount Pinatubo is on the Central Luzon landmass between the South China Sea and the Philippine Sea. The 1991 eruption of Pinatubo was andesite and dacite in the pre-climactic phase but only dacite in the climactic phase. The climactic phase had a volume of 3.7-5.3 km^3. The eruption consisted of sequentially increasing ash emissions, dome growth, 4 vertical eruptions with continued dome growth, 13 pyroclastic flows and a climactic vertical eruption with associated pyroclastic flows. The pre-climactic phase was phreatomagmatic.

Lake Taupo

The Hatepe eruption in 232+/-12 AD was the latest major eruption at Lake Taupo in New Zealand's Taupo Volcanic Zone. There was minor initial phreatomagmatic activity followed by the dry venting of 6 km^3 of rhyolite forming the Hatepe Plinian Pumice. The vent was then infiltrated by large amounts of water causing the phreatomagmatic eruption that deposited the 2.5 km3 Hatepe Ash. The water eventually stopped the eruption though large amounts of water were still erupted from the vent. The eruption resumed with phreatomagmatic activity that deposited the Rotongaio Ash.

Phreatic Eruption

A phreatic eruption, also called a phreatic explosion, ultravulcanian eruption or steam-blast eruption, occurs when magma heats ground or surface water. The extreme temperature of the magma (anywhere from 500 to 1,170 °C (932 to 2,138 °F)) causes near-instantaneous evaporation to steam, resulting in an explosion of steam, water, ash, rock, and volcanic bombs. At Mount St. Helens, hundreds of steam explosions preceded a 1980 plinian eruption of the volcano. A less intense geothermal event may result in a mud volcano.

Phreatic eruption at the summit of Mount St. Helens, Washington, in the spring of 1980

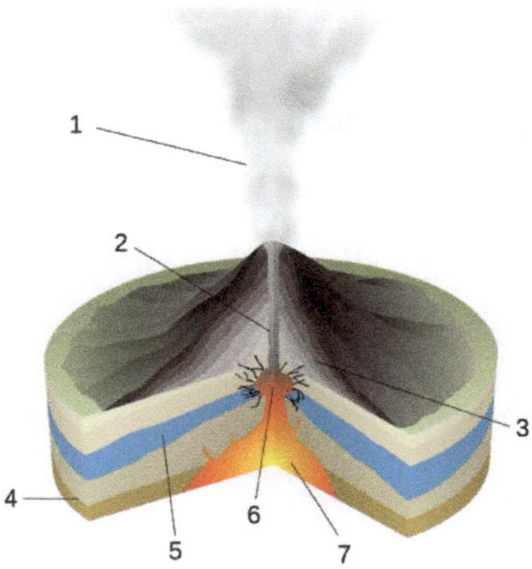

A scheme of a phreatic eruption

Phreatic eruptions typically include steam and rock fragments; the inclusion of lava is unusual. The temperature of the fragments can range from cold to incandescent. If molten magma is included, it is classified as a phreatomagmatic eruption. These eruptions occasionally create broad, low-relief craters called *maars*. Phreatic explosions can be accompanied by carbon dioxide or hydrogen sulfide gas emissions. The former can asphyxiate at sufficient concentration; the latter is a broad spectrum poison. A 1979 phreatic eruption on the island of Java killed 140 people, most of whom were overcome by poisonous gases.

Phreatic eruptions are classed as volcanic eruptions because a phreatic eruption could bring juvenile material to the surface.

It is believed that the 1883 eruption of Krakatoa, which obliterated most of the volcanic island and created the loudest sound in recorded history, was a phreatic event. Kilauea, in Hawaii, has a long record of phreatic explosions; a 1924 phreatic eruption hurled rocks estimated at eight tons up to a distance of one kilometer. Additional examples are the 1963–65 eruption of Surtsey, the 1965 eruption of Taal Volcano, the 1982 Mount Tarumae eruption, and the 2014 eruption of Mount Ontake.

Vulcanian Eruption

The term *vulcanian* was first used by Giuseppe Mercalli, witnessing the 1888-1890 eruptions on the island of Vulcano. His description of the eruption style is now used all over the world for eruptions characterised by a dense cloud of ash-laden gas exploding from the crater and rising high above the peak. Mercalli described vulcanian eruptions as "...Explosions like cannon fire at irregular intervals..." Their explosive nature is due to increased silica content of the magma. Almost all types of magma can be involved, but magma with about 55% or more silica (e.g. basaltic andesite) is most common.

Increasing silica levels increase the viscosity of the magma which means increased explosiveness. They usually commence with phreatomagmatic eruptions which can be extremely noisy due the rising magma heating water in the ground. This is usually followed by the explosive clearing of the vent and the eruption column is dirty grey to black as old weathered rocks are blasted out of the vent. As the vent clears, further ash clouds become grey-white and creamy in colour, with convolutions of the ash similar to those of plinian eruptions.

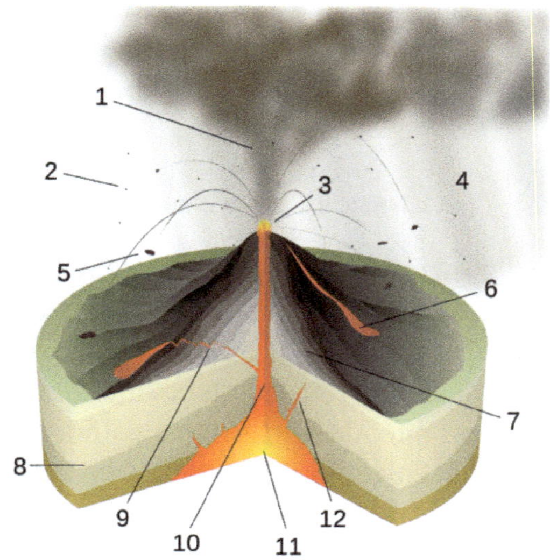

Volcanic eruption: 1 Ash plume, 2 Lapilli, 3 Lava fountain, 4 Volcanic ash rain, 5 Volcanic bomb, 6 Lava flow, 7 Layers of lava and ash, 8 Stratum, 9 Sill, 10 Magma conduit, 11 Magma chamber, 12 Dike

Characteristics

Vulcanian eruptions display several common characteristics. The mass of rock ejected during the eruption is usually between 10^2 - 10^6 tonnes and contains a high proportion of non-juvinial material (> 50%). During active periods of volcanic activity, intervals between explosions vary from less than 1 minute (e.g. Anak Krakatoa) to about a day. Pyroclastic flows are also common features of this type of eruption. The gas streaming phase of vulcanian eruptions are characterised by discrete cannon-like explosions. These expulsions of gas can reach supersonic velocities resulting in shock waves.

The tephra is dispersed over a wider area than that from Strombolian eruptions. The pyroclastic rock and the base surge deposits form an ash volcanic cone, while the ash covers a large surrounding area. The eruption ends with a flow of viscous lava. Vulcanian eruptions may throw large metre-size blocks several hundred metres, occasionally up to several kilometres.

Vulcanian eruptions are dangerous to persons within several hundred metres of the vent. Volcanic bombs are common products of this type of eruption. These are initially molten blobs of lava, which rapidly cool into blocks often 2 to 3 m across. At Galeras, a vulcanian eruption ejected bombs which struck several volcanologists who were in the crater, some of whom died or suffered severe injuries.

The Gran Cratere, Vulcano. A sense of scale is provided by the tourist visible near the centre of the crater.

1930 Eruption of Stromboli

The 11 September 1930 eruption of Stromboli was a vulcanian eruption. It started at 08:10 hours (local), when ash was vented for about 10 minutes. Then at 09:52 two incredibly powerful explosions occurred which shook the whole island. Blocks were hurled about 2 km. These fell out of the sky smashing through buildings etc. A tsunami 2 to 2.5 m high was generated. By 10:40 the explosive phase of the eruption was over. Expulsion of lava followed, this flowed down the Sciara del Fuoco, lasting into the night. At the same time incandescent scoria flowed down the Vallonazzo Valley and entered the sea near Piscità.

It is believed that water entered due to a partial collapse of the conduit. The water flashed into steam and took the easiest "escape route" via the open conduit. Expansion by contact with the molten magma generated the two very large explosions.

There were six deaths. Four fishermen died at sea when the avalanches of hot scoria caused the sea to become very disturbed. One person was killed in Stromboli village by falling blocks, and the sixth was killed by the tsunami.

Plinian Eruption

Plinian eruptions, also known as Vesuvian eruptions, are volcanic eruptions marked by their similarity to the eruption of Mount Vesuvius in 79 CE. The eruption was described in a letter written by Pliny the Younger; it killed his uncle, Pliny the Elder.

Plinian eruptions are marked by columns of gas and volcanic ash extending high into the stratosphere, the second layer of Earth's atmosphere. The key characteristics are ejection of large amount of pumice and very powerful continuous gas blast eruptions.

According to the Volcanic Explosivity Index, Plinian eruptions have a VEI of 4, 5 or 6.

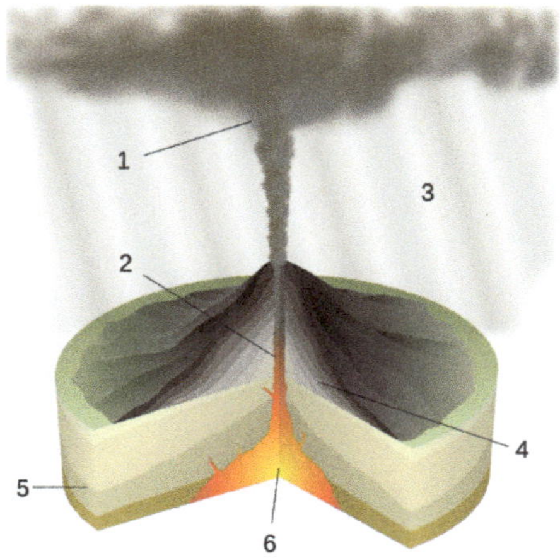

Plinian eruption: 1: ash plume, 2: magma conduit, 3: volcanic ash rain, 4: layers of lava and ash, |5: stratum, 6: magma chamber

Short eruptions can end in less than a day, but longer events can take several days to months. The longer eruptions begin with production of clouds of volcanic ash, sometimes with pyroclastic flows. The amount of magma erupted can be so large that it depletes the magma chamber below, causing the top of the volcano to collapse, resulting in a caldera. Fine ash can deposit over large areas. Plinian eruptions are often accompanied by loud noises, such as those generated by Krakatoa.

The lava is usually rhyolitic and rich in silicates. Basaltic lavas are unusual for Plinian eruptions; the most recent example is the 1886 eruption of Mount Tarawera on New Zealand's North Island.

Pliny's Description

A Stone Pine, the type of tree used by Pliny to describe the eruption.

Pliny described his uncle's involvement from the first observation of the eruption:

On the 24th of August, about one in the afternoon, my mother desired him to observe a cloud which appeared of a very unusual size and shape. He had just taken a turn in the sun and, after bathing himself in cold water, and making a light luncheon, gone back to his books: he immediately arose and went out upon a rising ground from whence he might get a better sight of this very uncommon appearance. A cloud, from which mountain was uncertain, at this distance (but it was found afterwards to come from Mount Vesuvius), was ascending, the appearance of which I cannot give you a more exact description of than by likening it to that of a pine tree, for it shot up to a great height in the form of a very tall trunk, which spread itself out at the top into a sort of branches; occasioned, I imagine, either by a sudden gust of air that impelled it, the force of which decreased as it advanced upwards, or the cloud itself being pressed back again by its own weight, expanded in the manner I have mentioned; it appeared sometimes bright and sometimes dark and spotted, according as it was either more or less impregnated with earth and cinders. This phenomenon seemed to a man of such learning and research as my uncle extraordinary and worth further looking into.

—*Sixth Book of Letters, Letter 16, translation by William Melmoth*

Pliny the Elder set out to rescue the victims from their perilous position on the shore of the Bay of Naples, and launched his galleys, crossing the bay to Stabiae (near the modern town of Castellammare di Stabia). Pliny the Younger provided an account of his death, and suggested that he collapsed and died through inhaling poisonous gases emitted from the volcano. His body was found interred under the ashes of the Vesuvius with no apparent injuries on 26 August, after the plume had dispersed, confirming asphyxiation or poisoning.

References

- Dosseto, A., Turner, S. P. and Van-Orman, J. A. (editors) (2011). "Timescales of Magmatic Processes: From Core to Atmosphere". Wiley-Blackwell. ISBN 978-1-4443-3260-5.
- Donald Hyndman & David Hyndman (April 2008). Natural Hazards and Disasters. Cengage Learning. pp. 134–135. ISBN 978-0-495-31667-1. Retrieved 2 August 2010.
- Alwyn Scarth & Jean-Claude Tanguy (31 May 2001). Volcanoes of Europe. Oxford University Press. p. 264. ISBN 978-0-19-521754-4. Retrieved 5 August 2010.
- Skinner, Brian J. (2004). Dynamic Earth: An Introduction to Physical Geology. John Wiley & Sons. Inc. Hoboken, NJ. ISBN 978-0-471-15228-6.
- Ruprecht P, Plank T. Feeding andesitic eruptions with a high-speed connection from the mantle. Nature. 2013;500(7460):68-72.
- "Volcanoes of Canada: Volcanic eruptions". Geological Survey of Canada. Natural Resources Canada. 2 April 2009. Retrieved 3 August 2010.
- Stephen Self; Jing-Xia Zhao; Rick E. Holasek; Ronnie C. Torres & Alan J. King. "The Atmospheric Impact of the 1991 Mount Pinatubo Eruption". USGS. Retrieved 3 August 2010.
- "X. Classification of Volcanic Eruptions: Surtseyan Eruptions". Lecture Notes. University of Alabama. Retrieved 5 August 2010.
- "Hunga Tonga-Hunga Ha'apai: Index of Monthly Reports". Global Volcanism Program. Smithsonian National Museum of Natural History. Retrieved 5 August 2010.

- Chadwick, Bill (10 January 2006). "Recent Submarine Volcanic Eruptions". Vents Program. NOAA. Retrieved 5 August 2010.
- "VHP Photo Glossary: Phreatic eruption". Volcano Hazards Program. U.S. Geological Survey. Retrieved 13 November 2010.
- Nelson, Stephan A. (30 September 2007). "Volcanoes, Magma, and Volcanic Eruptions". Tulane University. Retrieved 2 August 2010.
- "Lamington: Photo Gallery". Global Volcanism Program. Smithsonian National Museum of Natural History. Retrieved 2 August 2010.

7

Allied Fields of Volcanology

The allied fields of volcanology such as geology, geophysics and geochemistry are examined in this chapter. The study of solid earth and the rocks it is composed of is geology while geophysics is the science concerned with the physical properties and processes of the Earth. All these fields are mainly concerned with the physical aspects and processes of our planet.

Geology

Geology ("earth" and "study of, discourse") is an earth science comprising the study of solid Earth, the rocks of which it is composed, and the processes by which they change. Geology can also refer generally to the study of the solid features of any celestial body (such as the geology of the Moon or Mars).

Geology gives insight into the history of the Earth by providing the primary evidence for plate tectonics, the evolutionary history of life, and past climates. Geology is important for mineral and hydrocarbon exploration and exploitation, evaluating water resources, understanding of natural hazards, the remediation of environmental problems, and for providing insights into past climate change. Geology also plays a role in geotechnical engineering and is a major academic discipline.

Geologic Materials

The majority of geological data comes from research on solid Earth materials. These typically fall into one of two categories: rock and unconsolidated material.

Rock

There are three major types of rock: igneous, sedimentary, and metamorphic. The rock cycle is an important concept in geology which illustrates the relationships between these three types of rock, and magma. When a rock crystallizes from melt (magma and/or lava), it is an igneous rock. This rock can be weathered and eroded, and then redeposited and lithified into a sedimentary rock, or be turned into a metamorphic rock due to heat and pressure that change the mineral content of the rock which gives it a characteristic fabric. The sedimentary rock can then be subsequently turned into a metamorphic rock due to heat and pressure and is then weathered, eroded, deposited, and lithified, ultimately becoming a sedimentary rock. Sedimentary rock may also be re-eroded and redeposited, and metamorphic rock may also undergo additional metamorphism. All three types of rocks may be re-melted; when this happens, a new magma is formed, from which an igneous rock may once again crystallize.

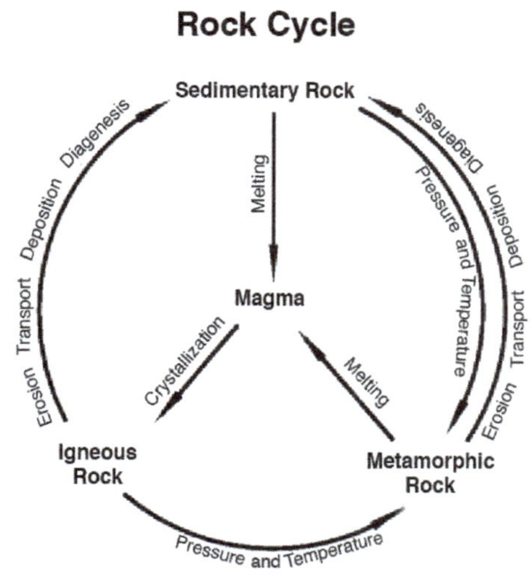

This schematic diagram of the rock cycle shows the relationship between magma and sedimentary, metamorphic, and igneous rock

The majority of research in geology is associated with the study of rock, as rock provides the primary record of the majority of the geologic history of the Earth.

Unconsolidated Material

Geologists also study unlithified material, which typically comes from more recent deposits. These materials are superficial deposits which lie above the bedrock. Because of this, the study of such material is often known as Quaternary geology, after the recent Quaternary Period. This includes the study of sediment and soils, including studies in geomorphology, sedimentology, and paleoclimatology.

Whole-Earth Structure

Plate Tectonics

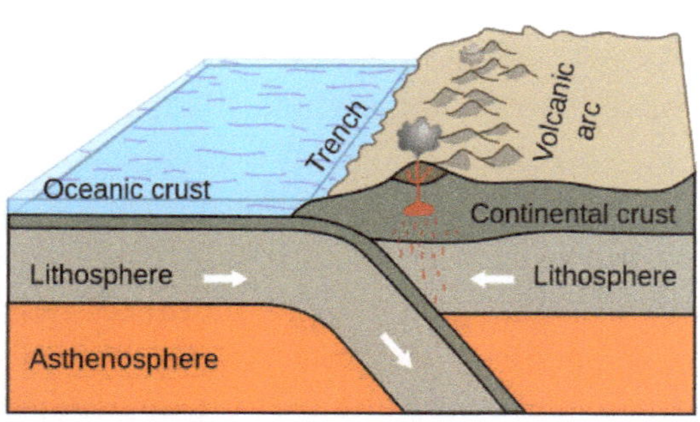

Oceanic-continental convergence resulting in subduction and volcanic arcs illustrates one effect of plate tectonics.

In the 1960s, a series of discoveries, the most important of which was seafloor spreading, showed that the Earth's lithosphere, which includes the crust and rigid uppermost portion of the upper mantle, is separated into a number of tectonic plates that move across the plastically deforming, solid, upper mantle, which is called the asthenosphere. There is an intimate coupling between the movement of the plates on the surface and the convection of the mantle: oceanic plate motions and mantle convection currents always move in the same direction, because the oceanic lithosphere is the rigid upper thermal boundary layer of the convecting mantle. This coupling between rigid plates moving on the surface of the Earth and the convecting mantle is called plate tectonics.

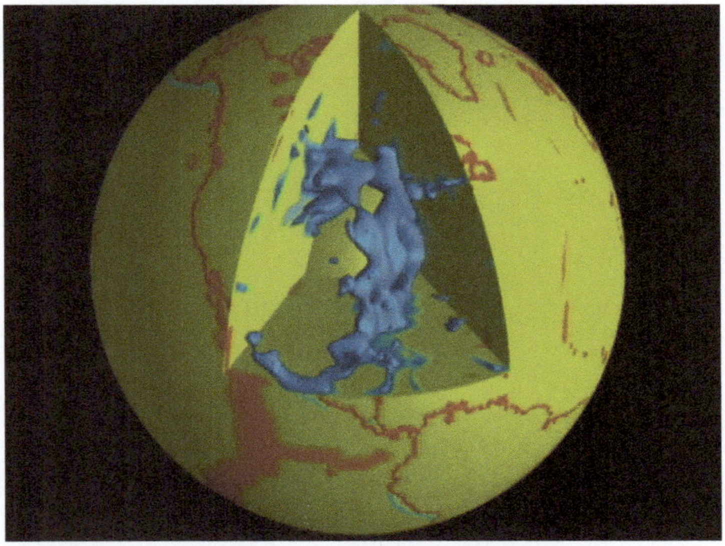

On this diagram, subducting slabs are in blue, and continental margins and a few plate boundaries are in red. The blue blob in the cutaway section is the seismically imaged Farallon Plate, which is subducting beneath North America. The remnants of this plate on the Surface of the Earth are the Juan de Fuca Plate and Explorer plate in the Northwestern USA / Southwestern Canada, and the Cocos Plate on the west coast of Mexico.

The development of plate tectonics provided a physical basis for many observations of the solid Earth. Long linear regions of geologic features could be explained as plate boundaries. Mid-ocean ridges, high regions on the seafloor where hydrothermal vents and volcanoes exist, were explained as divergent boundaries, where two plates move apart. Arcs of volcanoes and earthquakes were explained as convergent boundaries, where one plate subducts under another. Transform boundaries, such as the San Andreas Fault system, resulted in widespread powerful earthquakes. Plate tectonics also provided a mechanism for Alfred Wegener's theory of continental drift, in which the continents move across the surface of the Earth over geologic time. They also provided a driving force for crustal deformation, and a new setting for the observations of structural geology. The power of the theory of plate tectonics lies in its ability to combine all of these observations into a single theory of how the lithosphere moves over the convecting mantle.

Earth Structure

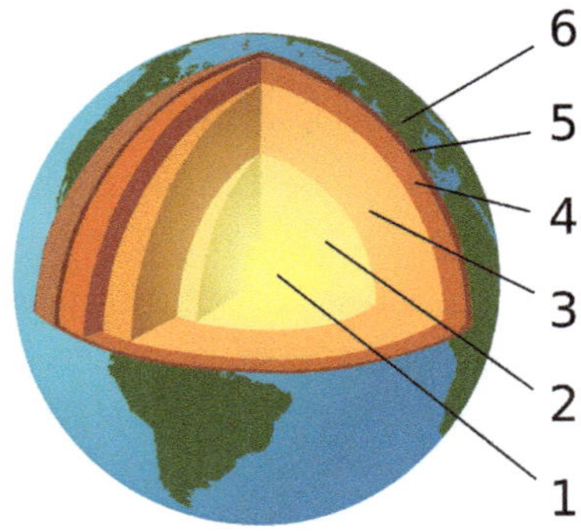

The Earth's layered structure. (1) inner core; (2) outer core; (3) lower mantle; (4) upper mantle; (5) lithosphere; (6) crust (part of the lithosphere)

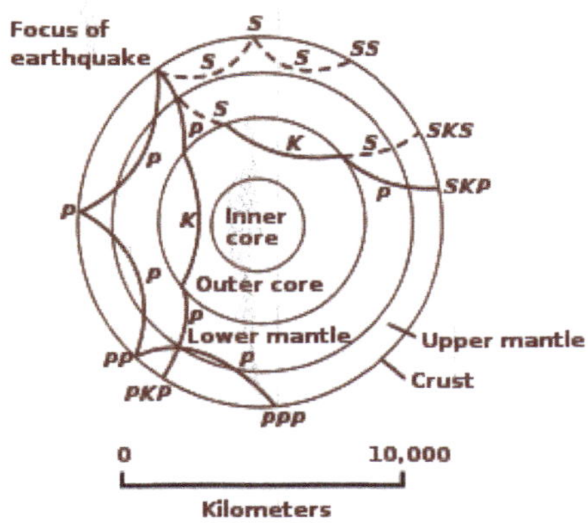

Earth layered structure. Typical wave paths from earthquakes like these gave early seismologists insights into the layered structure of the Earth

Advances in seismology, computer modeling, and mineralogy and crystallography at high temperatures and pressures give insights into the internal composition and structure of the Earth.

Seismologists can use the arrival times of seismic waves in reverse to image the interior of the Earth. Early advances in this field showed the existence of a liquid outer core (where shear waves were not able to propagate) and a dense solid inner core. These advances led to the development of a layered model of the Earth, with a crust and lithosphere on top, the mantle below (separated within itself by seismic discontinuities at 410 and 660 kilometers), and the outer core and inner core below that. More recently, seismologists have been able to create detailed images of wave speeds inside the earth in the same way a doctor images a body in a CT scan. These images have led to a much more detailed view of the interior of the Earth, and have replaced the simplified layered model with a much more dynamic model.

Mineralogists have been able to use the pressure and temperature data from the seismic and modelling studies alongside knowledge of the elemental composition of the

Earth to reproduce these conditions in experimental settings and measure changes in crystal structure. These studies explain the chemical changes associated with the major seismic discontinuities in the mantle and show the crystallographic structures expected in the inner core of the Earth.

Geologic Time

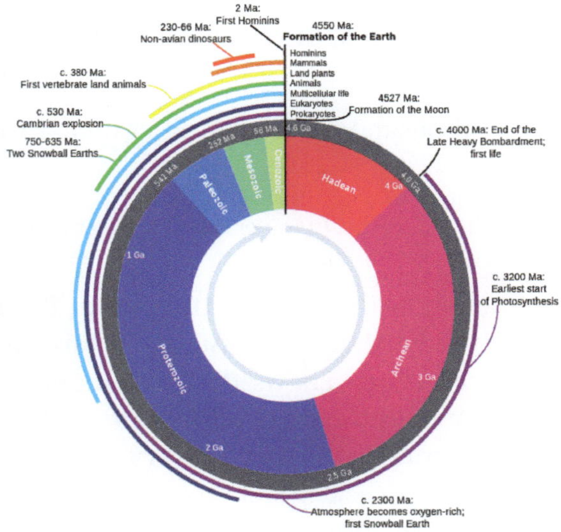

Geological time put in a diagram called a geological clock, showing the relative lengths of the eons of the Earth's history.

The geologic time scale encompasses the history of the Earth. It is bracketed at the old end by the dates of the earliest Solar System material at 4.567 Ga, (gigaannum: billion years ago) and the age of the Earth at 4.54 Ga at the beginning of the informally recognized Hadean eon. At the young end of the scale, it is bracketed by the present day in the Holocene epoch.

Important Milestones

- 4.567 Ga: Solar system formation
- 4.54 Ga: Accretion of Earth
- c. 4 Ga: End of Late Heavy Bombardment, first life
- c. 3.5 Ga: Start of photosynthesis
- c. 2.3 Ga: Oxygenated atmosphere, first snowball Earth
- 730–635 Ma (megaannum: million years ago): second snowball Earth
- 542 ± 0.3 Ma: Cambrian explosion – vast multiplication of hard-bodied life; first abundant fossils; start of the Paleozoic
- c. 380 Ma: First vertebrate land animals

- 250 Ma: Permian-Triassic extinction – 90% of all land animals die; end of Paleozoic and beginning of Mesozoic
- 66 Ma: Cretaceous–Paleogene extinction – Dinosaurs die; end of Mesozoic and beginning of Cenozoic
- c. 7 Ma: First hominins appear
- 3.9 Ma: First Australopithecus, direct ancestor to modern Homo sapiens, appear
- 200 ka (kiloannum: thousand years ago): First modern Homo sapiens appear in East Africa

Brief Time Scale

The following four timelines show the geologic time scale. The first shows the entire time from the formation of the Earth to the present, but this compresses the most recent eon. Therefore, the second scale shows the most recent eon with an expanded scale. The second scale compresses the most recent era, so the most recent era is expanded in the third scale. Since the Quaternary is a very short period with short epochs, it is further expanded in the fourth scale. The second, third, and fourth timelines are therefore each subsections of their preceding timeline as indicated by asterisks. The Holocene (the latest epoch) is too small to be shown clearly on the third timeline on the right, another reason for expanding the fourth scale. The Pleistocene (P) epoch. Q stands for the Quaternary period.

Dating Methods

Geologists use a variety of methods to give both relative and absolute dates to geological events. They then use these dates to find the rates at which processes occur.

Relative Dating

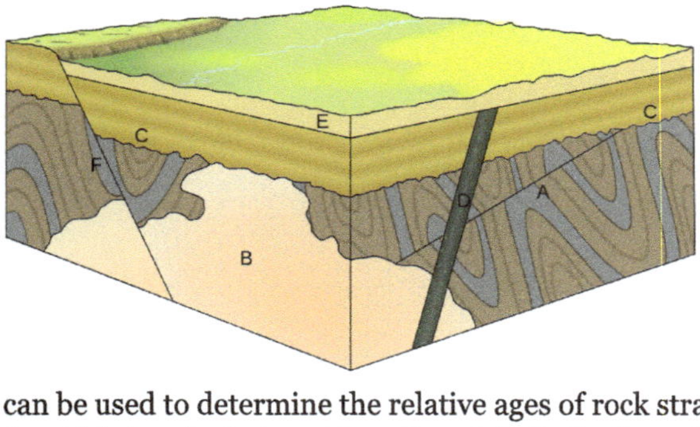

Cross-cutting relations can be used to determine the relative ages of rock strata and other geological structures. Explanations: A – folded rock strata cut by a thrust fault; B – large intrusion (cutting through A); C – erosional angular unconformity (cutting off A & B) on which rock strata were deposited; D – volcanic dyke (cutting through A, B & C); E – even younger rock strata (overlying C & D); F – normal fault (cutting through A, B, C & E).

Methods for relative dating were developed when geology first emerged as a formal science. Geologists still use the following principles today as a means to provide information about geologic history and the timing of geologic events.

The principle of Uniformitarianism states that the geologic processes observed in operation that modify the Earth's crust at present have worked in much the same way over geologic time. A fundamental principle of geology advanced by the 18th century Scottish physician and geologist James Hutton, is that "the present is the key to the past." In Hutton's words: "the past history of our globe must be explained by what can be seen to be happening now."

The principle of intrusive relationships concerns crosscutting intrusions. In geology, when an igneous intrusion cuts across a formation of sedimentary rock, it can be determined that the igneous intrusion is younger than the sedimentary rock. There are a number of different types of intrusions, including stocks, laccoliths, batholiths, sills and dikes.

The principle of cross-cutting relationships pertains to the formation of faults and the age of the sequences through which they cut. Faults are younger than the rocks they cut; accordingly, if a fault is found that penetrates some formations but not those on top of it, then the formations that were cut are older than the fault, and the ones that are not cut must be younger than the fault. Finding the key bed in these situations may help determine whether the fault is a normal fault or a thrust fault.

The principle of inclusions and components states that, with sedimentary rocks, if inclusions (or *clasts*) are found in a formation, then the inclusions must be older than the formation that contains them. For example, in sedimentary rocks, it is common for gravel from an older formation to be ripped up and included in a newer layer. A similar situation with igneous rocks occurs when xenoliths are found. These foreign bodies are picked up as magma or lava flows, and are incorporated, later to cool in the matrix. As a result, xenoliths are older than the rock which contains them.

The Permian through Jurassic stratigraphy of the Colorado Plateau area of southeastern Utah is a great example of both Original Horizontality and the Law of Superposition.

These strata make up much of the famous prominent rock formations in widely spaced protected areas such as Capitol Reef National Park and Canyonlands National Park. From top to bottom: Rounded tan domes of the Navajo Sandstone, layered red Kayenta Formation, cliff-forming, vertically jointed, red Wingate Sandstone, slope-forming, purplish Chinle Formation, layered, lighter-red Moenkopi Formation, and white, layered Cutler Formation sandstone. Picture from Glen Canyon National Recreation Area, Utah.

The principle of original horizontality states that the deposition of sediments occurs as essentially horizontal beds. Observation of modern marine and non-marine sediments in a wide variety of environments supports this generalization (although cross-bedding is inclined, the overall orientation of cross-bedded units is horizontal).

The principle of superposition states that a sedimentary rock layer in a tectonically undisturbed sequence is younger than the one beneath it and older than the one above it. Logically a younger layer cannot slip beneath a layer previously deposited. This principle allows sedimentary layers to be viewed as a form of vertical time line, a partial or complete record of the time elapsed from deposition of the lowest layer to deposition of the highest bed.

The principle of faunal succession is based on the appearance of fossils in sedimentary rocks. As organisms exist at the same time period throughout the world, their presence or (sometimes) absence may be used to provide a relative age of the formations in which they are found. Based on principles laid out by William Smith almost a hundred years before the publication of Charles Darwin's theory of evolution, the principles of succession were developed independently of evolutionary thought. The principle becomes quite complex, however, given the uncertainties of fossilization, the localization of fossil types due to lateral changes in habitat (facies change in sedimentary strata), and that not all fossils may be found globally at the same time.

Absolute Dating

Geologists also use methods to determine the absolute age of rock samples and geological events. These dates are useful on their own and may also be used in conjunction with relative dating methods or to calibrate relative methods.

At the beginning of the 20th century, important advancement in geological science was facilitated by the ability to obtain accurate absolute dates to geologic events using radioactive isotopes and other methods. This changed the understanding of geologic time. Previously, geologists could only use fossils and stratigraphic correlation to date sections of rock relative to one another. With isotopic dates it became possible to assign absolute ages to rock units, and these absolute dates could be applied to fossil sequences in which there was datable material, converting the old relative ages into new absolute ages.

For many geologic applications, isotope ratios of radioactive elements are measured in minerals that give the amount of time that has passed since a rock passed through its particular closure temperature, the point at which different radiometric isotopes stop diffusing into and out of the crystal lattice. These are used in geochronologic and thermochronologic studies. Common methods include uranium-lead dating, potassium-argon dating, argon-argon dating and uranium-thorium dating. These methods are used for a variety of applications. Dating of lava and volcanic ash layers found within a stratigraphic sequence can provide absolute age data for sedimentary rock units which do not contain radioactive isotopes and calibrate relative dating techniques. These methods can also be used to determine ages of pluton emplacement. Thermochemical techniques can be used to determine temperature profiles within the crust, the uplift of mountain ranges, and paleotopography.

Fractionation of the lanthanide series elements is used to compute ages since rocks were removed from the mantle.

Other methods are used for more recent events. Optically stimulated luminescence and cosmogenic radionucleide dating are used to date surfaces and/or erosion rates. Dendrochronology can also be used for the dating of landscapes. Radiocarbon dating is used for geologically young materials containing organic carbon.

Geological Development of an Area

The geology of an area changes through time as rock units are deposited and inserted and deformational processes change their shapes and locations.

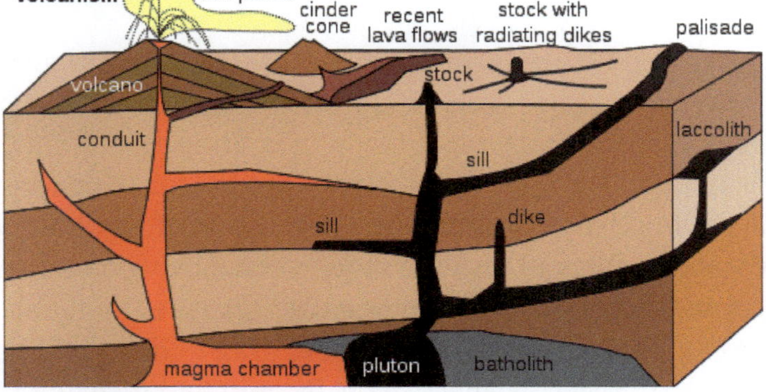

An originally horizontal sequence of sedimentary rocks (in shades of tan) are affected by igneous activity. Deep below the surface are a magma chamber and large associated igneous bodies. The magma chamber feeds the volcano, and sends offshoots of magma that will later crystallize into dikes and sills. Magma also advances upwards to form intrusive igneous bodies. The diagram illustrates both a cinder cone volcano, which releases ash, and a composite volcano, which releases both lava and ash.

Rock units are first emplaced either by deposition onto the surface or intrusion into the overlying rock. Deposition can occur when sediments settle onto the surface of the Earth and later lithify into sedimentary rock, or when as volcanic material such as volcanic ash or lava flows blanket the surface. Igneous intrusions such as batholiths, laccoliths, dikes, and sills, push upwards into the overlying rock, and crystallize as they intrude.

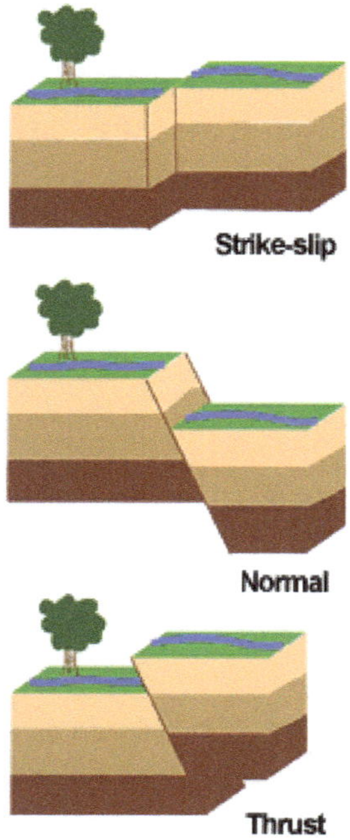

An illustration of the three types of faults. Strike-slip faults occur when rock units slide past one another, normal faults occur when rocks are undergoing horizontal extension, and thrust faults occur when rocks are undergoing horizontal shortening.

After the initial sequence of rocks has been deposited, the rock units can be deformed and/or metamorphosed. Deformation typically occurs as a result of horizontal shortening, horizontal extension, or side-to-side (strike-slip) motion. These structural regimes broadly relate to convergent boundaries, divergent boundaries, and transform boundaries, respectively, between tectonic plates.

When rock units are placed under horizontal compression, they shorten and become thicker. Because rock units, other than muds, do not significantly change in volume, this is accomplished in two primary ways: through faulting and folding. In the shallow crust, where brittle deformation can occur, thrust faults form, which cause deeper rock to move on top of shallower rock. Because deeper rock is often older, as noted by the principle of superposition, this can result in older rocks moving on top of younger ones. Movement along faults can result in folding, either because the faults are not planar or because rock layers are dragged along, forming drag folds as slip occurs along the fault. Deeper in the Earth, rocks behave plastically, and fold instead of faulting. These folds can either be those where the material in the center of the fold buckles upwards, creating "antiforms", or where it buckles downwards, creating "synforms". If the tops of the rock units within the folds remain pointing upwards, they are called anticlines and synclines, respectively. If some of the units in the fold are facing downward, the structure is called an overturned anticline or syncline, and if all of the rock units are overturned or the correct up-direction is unknown, they are simply called by the most general terms, antiforms and synforms.

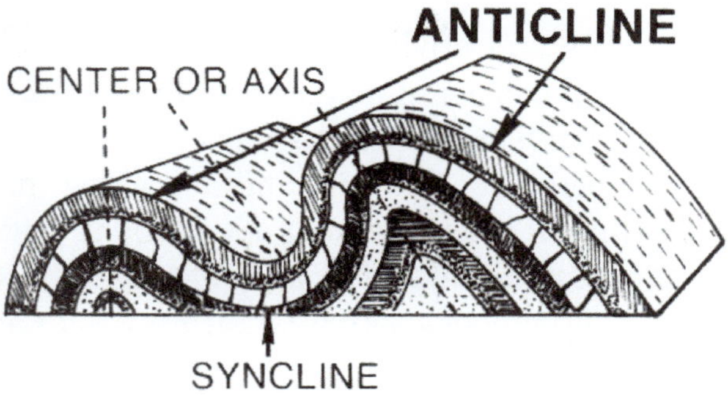

A diagram of folds, indicating an anticline and a syncline.

Even higher pressures and temperatures during horizontal shortening can cause both folding and metamorphism of the rocks. This metamorphism causes changes in the mineral composition of the rocks; creates a foliation, or planar surface, that is related to mineral growth under stress. This can remove signs of the original textures of the rocks, such as bedding in sedimentary rocks, flow features of lavas, and crystal patterns in crystalline rocks.

Extension causes the rock units as a whole to become longer and thinner. This is primarily accomplished through normal faulting and through the ductile stretching and thinning. Normal faults drop rock units that are higher below those that are lower. This typically results in younger units being placed below older units. Stretching of units can result in their thinning; in fact, there is a location within the Maria Fold and Thrust Belt in which the entire sedimentary sequence of the Grand Canyon can be seen over a length of less than a meter. Rocks at the depth to be ductilely stretched are often also metamorphosed. These stretched rocks can also pinch into lenses, known as boudins, after the French word for "sausage", because of their visual similarity.

Where rock units slide past one another, strike-slip faults develop in shallow regions, and become shear zones at deeper depths where the rocks deform ductilely.

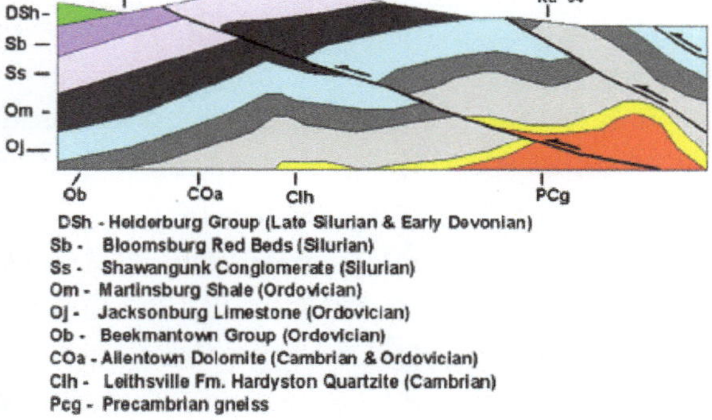

Geologic cross section of Kittatinny Mountain. This cross section shows metamorphic rocks, overlain by younger sediments deposited after the metamorphic event. These rock units were later folded and faulted during the uplift of the mountain.

The addition of new rock units, both depositionally and intrusively, often occurs during deformation. Faulting and other deformational processes result in the creation of topographic gradients, causing material on the rock unit that is increasing in elevation to be eroded by hillslopes and channels. These sediments are deposited on the rock unit that is going down. Continual motion along the fault maintains the topographic gradient in spite of the movement of sediment, and continues to create accommodation space for the material to deposit. Deformational events are often also associated with volcanism and igneous activity. Volcanic ashes and lavas accumulate on the surface, and igneous intrusions enter from below. Dikes, long, planar igneous intrusions, enter along cracks, and therefore often form in large numbers in areas that are being actively deformed. This can result in the emplacement of dike swarms, such as those that are observable across the Canadian shield, or rings of dikes around the lava tube of a volcano.

All of these processes do not necessarily occur in a single environment, and do not necessarily occur in a single order. The Hawaiian Islands, for example, consist almost entirely of layered basaltic lava flows. The sedimentary sequences of the mid-continental United States and the Grand Canyon in the southwestern United States contain almost-undeformed stacks of sedimentary rocks that have remained in place since Cambrian time. Other areas are much more geologically complex. In the southwestern United States, sedimentary, volcanic, and intrusive rocks have been metamorphosed, faulted, foliated, and folded. Even older rocks, such as the Acasta gneiss of the Slave craton in northwestern Canada, the oldest known rock in the world have been metamorphosed to the point where their origin is undiscernable without laboratory analysis. In addition, these processes can occur in stages. In many places, the Grand Canyon in the southwestern United States being a very visible example, the lower rock units were metamorphosed and deformed, and then deformation ended and the upper, undeformed units were deposited. Although any amount of rock emplacement and rock deformation can occur, and they can occur any number of times, these concepts provide a guide to understanding the geological history of an area.

Methods of Geology

Geologists use a number of field, laboratory, and numerical modeling methods to decipher Earth history and to understand the processes that occur on and inside the Earth. In typical geological investigations, geologists use primary information related to petrology (the study of rocks), stratigraphy (the study of sedimentary layers), and structural geology (the study of positions of rock units and their deformation). In many cases, geologists also study modern soils, rivers, landscapes, and glaciers; investigate past and current life and biogeochemical pathways, and use geophysical methods to investigate the subsurface. Sub-specialities of geology may distinguish endogenous and exogenous geology.

Field Methods

A standard Brunton Pocket Transit, commonly used by geologists for mapping and surveying.

A typical USGS field mapping camp in the 1950s

Geological field work varies depending on the task at hand. Typical fieldwork could consist of:

- Geological mapping
    - Structural mapping: identifying the locations of major rock units and the faults and folds that led to their placement there.
    - Stratigraphic mapping: pinpointing the locations of sedimentary facies (lithofacies and biofacies) or the mapping of isopachs of equal thickness of sedimentary rock
    - Surficial mapping: recording the locations of soils and surficial deposits
- Surveying of topographic features
    - compilation of topographic maps
    - Work to understand change across landscapes, including:
        - Patterns of erosion and deposition
        - River-channel change through migration and avulsion
        - Hillslope processes
- Subsurface mapping through geophysical methods
    - These methods include:

        - Shallow seismic surveys
        - Ground-penetrating radar
        - Aeromagnetic surveys
        - Electrical resistivity tomography
    - They aid in:
        - Hydrocarbon exploration
        - Finding groundwater
        - Locating buried archaeological artifacts
- High-resolution stratigraphy
    - Measuring and describing stratigraphic sections on the surface
    - Well drilling and logging
- Biogeochemistry and geomicrobiology
    - Collecting samples to:
        - determine biochemical pathways
        - identify new species of organisms
        - identify new chemical compounds
    - and to use these discoveries to:
        - understand early life on Earth and how it functioned and metabolized
        - find important compounds for use in pharmaceuticals
- Paleontology: excavation of fossil material
    - For research into past life and evolution
    - For museums and education
- Collection of samples for geochronology and thermochronology
- Glaciology: measurement of characteristics of glaciers and their motion

Petrology

In addition to identifying rocks in the field, petrologists identify rock samples in the laboratory. Two of the primary methods for identifying rocks in the laboratory are through optical microscopy and by using an electron microprobe. In an optical mineralogy analysis, petrologists analyze thin sections of rock samples using a petrographic microscope, where the minerals can be identified through their different properties in plane-polarized and cross-polarized light, including their birefringence, pleochroism, twinning, and interference properties with a conoscopic lens. In the electron microprobe, individual locations are analyzed for their exact chemical compositions and variation in composition within individual crystals. Stable and radioactive isotope studies provide insight into the geochemical evolution of rock units.

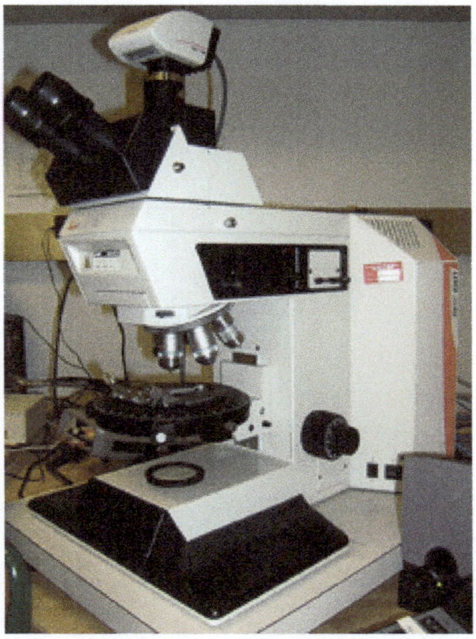

A petrographic microscope - an optical microscope fitted with cross-polarizing lenses, a conoscopic lens, and compensators (plates of anisotropic materials; gypsum plates and quartz wedges are common), for crystallographic analysis.

Petrologists can also use fluid inclusion data and perform high temperature and pressure physical experiments to understand the temperatures and pressures at which different mineral phases appear, and how they change through igneous and metamorphic processes. This research can be extrapolated to the field to understand metamorphic processes and the conditions of crystallization of igneous rocks. This work can also help to explain processes that occur within the Earth, such as subduction and magma chamber evolution.

Structural Geology

Structural geologists use microscopic analysis of oriented thin sections of geologic samples to observe the fabric within the rocks which gives information about strain within the crystalline structure of the rocks. They also plot and combine measurements of geological structures in order to better understand the orientations of faults and folds in order to reconstruct the history of rock deformation in the area. In addition, they perform analog and numerical experiments of rock deformation in large and small settings.

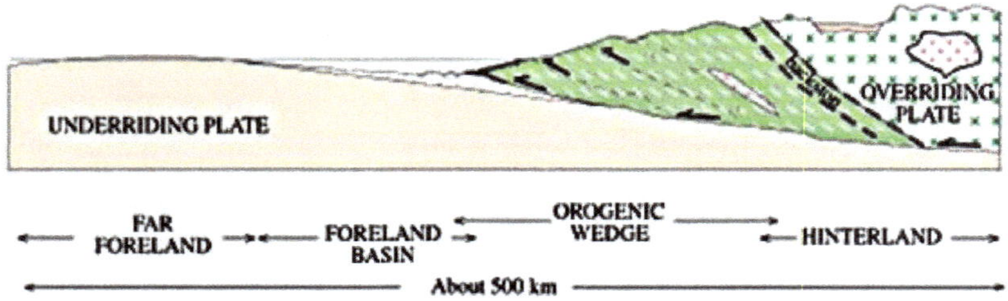

A diagram of an orogenic wedge. The wedge grows through faulting in the interior and along the main basal fault, called the décollement. It builds its shape into a critical taper, in which the angles within the wedge remain the same as failures inside the material balance failures along the décollement. It is analogous to a bulldozer pushing a pile of dirt, where the bulldozer is the overriding plate.

The analysis of structures is often accomplished by plotting the orientations of various features onto stereonets. A stereonet is a stereographic projection of a sphere onto a plane, in which planes are projected as lines and lines are projected as points. These can be used to find the locations of fold axes, relationships between faults, and relationships between other geologic structures.

Among the most well-known experiments in structural geology are those involving orogenic wedges, which are zones in which mountains are built along convergent tectonic plate boundaries. In the analog versions of these experiments, horizontal layers of sand are pulled along a lower surface into a back stop, which results in realistic-looking patterns of faulting and the growth of a critically tapered (all angles remain the same) orogenic wedge. Numerical models work in the same way as these analog models, though they are often more sophisticated and can include patterns of erosion and uplift in the mountain belt. This helps to show the relationship between erosion and the shape of a mountain range. These studies can also give useful information about pathways for metamorphism through pressure, temperature, space, and time.

Stratigraphy

In the laboratory, stratigraphers analyze samples of stratigraphic sections that can be returned from the field, such as those from drill cores. Stratigraphers also analyze data from geophysical surveys that show the locations of stratigraphic units in the subsurface. Geophysical data and well logs can be combined to produce a better view of the subsurface, and stratigraphers often use computer programs to do this in three dimensions. Stratigraphers can then use these data to reconstruct ancient processes occurring on the surface of the Earth, interpret past environments, and locate areas for water, coal, and hydrocarbon extraction.

In the laboratory, biostratigraphers analyze rock samples from outcrop and drill cores for the fossils found in them. These fossils help scientists to date the core and to understand the depositional environment in which the rock units formed. Geochronologists precisely date rocks within the stratigraphic section in order to provide better absolute bounds on the timing and rates of deposition. Magnetic stratigraphers look for signs of magnetic reversals in igneous rock units within the drill cores. Other scientists perform stable-isotope studies on the rocks to gain information about past climate.

Planetary Geology

Surface of Mars as photographed by the Viking 2 lander December 9, 1977.

With the advent of space exploration in the twentieth century, geologists have begun to look at other planetary bodies in the same ways that have been developed to study the Earth. This new field of study is called planetary geology (sometimes known as astrogeology) and relies on known geologic principles to study other bodies of the solar system.

Although the Greek-language-origin prefix *geo* refers to Earth, "geology" is often used in conjunction with the names of other planetary bodies when describing their composition and internal processes: examples are "the geology of Mars" and "Lunar geology". Specialised terms such as *selenology* (studies of the Moon), *areology* (of Mars), etc., are also in use.

Although planetary geologists are interested in studying all aspects of other planets, a significant focus is to search for evidence of past or present life on other worlds. This has led to many missions whose primary or ancillary purpose is to examine planetary bodies for evidence of life. One of these is the Phoenix lander, which analyzed Martian polar soil for water, chemical, and mineralogical constituents related to biological processes.

Applied Geology

Economic Geology

Economic geology is an important branch of geology which deals with different aspects of economic minerals being used by humankind to fulfill its various needs. The economic minerals are those which can be extracted profitably. Economic geologists help locate and manage the Earth's natural resources, such as petroleum and coal, as well as mineral resources, which include metals such as iron, copper, and uranium.

Mining Geology

Mining geology consists of the extractions of mineral resources from the Earth. Some resources of economic interests include gemstones, metals, and many minerals such as asbestos, perlite, mica, phosphates, zeolites, clay, pumice, quartz, and silica, as well as elements such as sulfur, chlorine, and helium.

Petroleum Geology

Petroleum geologists study the locations of the subsurface of the Earth which can contain extractable hydrocarbons, especially petroleum and natural gas. Because many of these reservoirs are found in sedimentary basins, they study the formation of these basins, as well as their sedimentary and tectonic evolution and the present-day positions of the rock units.

Mud log in process, a common way to study the lithology when drilling oil wells.

Engineering Geology

Engineering geology is the application of the geologic principles to engineering practice for the purpose of assuring that the geologic factors affecting the location, design, construction, operation, and maintenance of engineering works are properly addressed.

In the field of civil engineering, geological principles and analyses are used in order to ascertain the mechanical principles of the material on which structures are built. This allows tunnels to be built without collapsing, bridges and skyscrapers to be built with sturdy foundations, and buildings to be built that will not settle in clay and mud.

Hydrology and Environmental Issues

Geology and geologic principles can be applied to various environmental problems such as stream restoration, the restoration of brownfields, and the understanding of the interaction between natural habitat and the geologic environment. Groundwater hydrology, or hydrogeology, is used to locate groundwater, which can often provide a ready supply of uncontaminated water and is especially important in arid regions, and to monitor the spread of contaminants in groundwater wells.

Geologists also obtain data through stratigraphy, boreholes, core samples, and ice cores. Ice cores and sediment cores are used to for paleoclimate reconstructions, which tell geologists about past and present temperature, precipitation, and sea level across the globe. These datasets are our primary source of information on global climate change outside of instrumental data.

Natural Hazards

Geologists and geophysicists study natural hazards in order to enact safe building codes and warning systems that are used to prevent loss of property and life. Examples of important natural hazards that are pertinent to geology (as opposed those that are mainly or only pertinent to meteorology) are:

Rockfall in the Grand Canyon

- Avalanches
- Earthquakes
- Floods
- Landslides and debris flows
- River channel migration and avulsion
- Liquefaction
- Sinkholes
- Subsidence

- Tsunamis
- Volcanoes

History of Geology

The study of the physical material of the Earth dates back at least to ancient Greece when Theophrastus (372–287 BCE) wrote the work *Peri Lithon* (*On Stones*). During the Roman period, Pliny the Elder wrote in detail of the many minerals and metals then in practical use – even correctly noting the origin of amber.

William Smith's geologic map of England, Wales, and southern Scotland. Completed in 1815, it was the second national-scale geologic map, and by far the most accurate of its time.

Some modern scholars, such as Fielding H. Garrison, are of the opinion that the origin of the science of geology can be traced to Persia after the Muslim conquests had come to an end. Abu al-Rayhan al-Biruni (973–1048 CE) was one of the earliest Persian geologists, whose works included the earliest writings on the geology of India, hypothesizing that the Indian subcontinent was once a sea. Drawing from Greek and Indian scientific literature that were not destroyed by the Muslim conquests, the Persian scholar Ibn Sina (Avicenna, 981–1037) proposed detailed explanations for the formation of mountains, the origin of earthquakes, and other topics central to modern geology, which provided an essential foundation for the later development of the science. In China, the polymath Shen Kuo (1031–1095) formulated a hypothesis for the process of land formation: based on his observation of fossil animal shells in a geological stratum in a mountain hundreds of miles from the ocean, he inferred that the land was formed by erosion of the mountains and by deposition of silt.

Nicolas Steno (1638–1686) is credited with the law of superposition, the principle of original horizontality, and the principle of lateral continuity: three defining principles of stratigraphy.

The word *geology* was first used by Ulisse Aldrovandi in 1603, then by Jean-André Deluc in 1778 and introduced as a fixed term by Horace-Bénédict de Saussure in 1779. The word is derived from the Greek γ, *gê*, meaning "earth" and λόγος, *logos*, meaning "speech". But according to another source, the word "geology" comes from a Norwegian, Mikkel Pedersøn Escholt (1600–1699), who was a priest and scholar. Escholt first used the definition in his book titled, *Geologica Norvegica* (1657).

William Smith (1769–1839) drew some of the first geological maps and began the process of ordering rock strata (layers) by examining the fossils contained in them.

James Hutton is often viewed as the first modern geologist. In 1785 he presented a paper entitled *Theory of the Earth* to the Royal Society of Edinburgh. In his paper, he explained his theory that the Earth must be much older than had previously been supposed in order to allow enough time for mountains to be eroded and for sediments to form new rocks at the bottom of the sea, which in turn were raised up to become dry land. Hutton published a two-volume version of his ideas in 1795 (Vol. 1, Vol. 2).

Scotsman James Hutton, father of modern geology

Followers of Hutton were known as *Plutonists* because they believed that some rocks were formed by *vulcanism*, which is the deposition of lava from volcanoes, as opposed to the *Neptunists*, led by Abraham Werner, who believed that all rocks had settled out of a large ocean whose level gradually dropped over time.

The first geological map of the U.S. was produced in 1809 by William Maclure. In 1807, Maclure commenced the self-imposed task of making a geological survey of the United States. Almost every state in the Union was traversed and mapped by him, the Allegheny Mountains being crossed and recrossed some 50 times. The results of his unaided labours were submitted to the American Philosophical Society in a memoir entitled *Observations on the Geology of the United States explanatory of a Geological Map*, and published in the *Society's Transactions*, together with the nation's first geological map. This antedates William Smith's geological map of England by six years, although it was constructed using a different classification of rocks.

Sir Charles Lyell first published his famous book, *Principles of Geology*, in 1830. This book, which influenced the thought of Charles Darwin, successfully promoted the doctrine of uniformitarianism. This theory states that slow geological processes have occurred throughout the Earth's history and are still occurring today. In contrast, catastrophism is the theory that Earth's features formed in single, catastrophic events and remained unchanged thereafter. Though Hutton believed in uniformitarianism, the idea was not widely accepted at the time.

Much of 19th-century geology revolved around the question of the Earth's exact age. Estimates varied from a few hundred thousand to billions of years. By the early 20th century, radiometric dating allowed the Earth's age to be estimated at two billion years. The awareness of this vast amount of time opened the door to new theories about the processes that shaped the planet.

Some of the most significant advances in 20th-century geology have been the development of the theory of plate tectonics in the 1960s and the refinement of estimates of the planet's age. Plate tectonics theory arose from two separate geological observations: seafloor spreading and continental drift. The theory revolutionized the Earth sciences. Today the Earth is known to be approximately 4.5 billion years old.

Geophysics

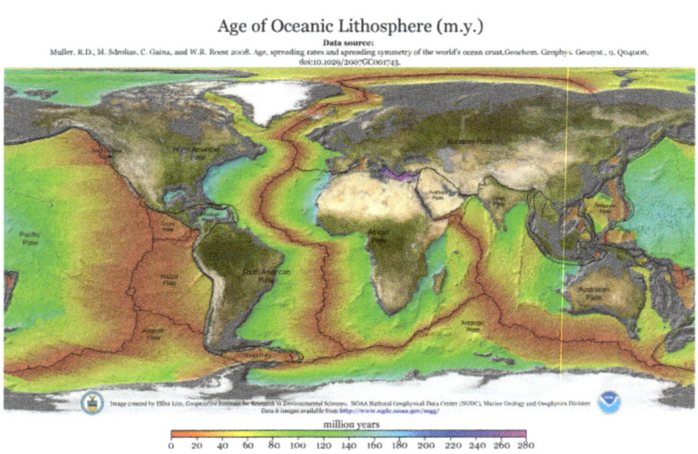

Age of the sea floor. Much of the dating information comes from magnetic anomalies.

Geophysics is a subject of natural science concerned with the physical processes and physical properties of the Earth and its surrounding space environment, and the use of quantitative methods for their analysis. The term *geophysics* sometimes refers only to the geological applications: Earth's shape; its gravitational and magnetic fields; its internal structure and composition; its dynamics and their surface expression in plate tectonics, the generation of magmas, volcanism and rock formation. However, modern geophysics organizations use a broader definition that includes the water cycle including snow and ice; fluid dynamics of the oceans and the atmosphere; electricity and magnetism in the ionosphere and magnetosphere and solar-terrestrial relations; and analogous problems associated with the Moon and other planets.

Although geophysics was only recognized as a separate discipline in the 19th century, its origins date back to ancient times. The first magnetic compasses were made from lodestones, while more modern magnetic compasses played an important role in the history of navigation. The first seismic instrument was built in 132 BC. Isaac Newton applied his theory of mechanics to the tides and the precession of the equinox; and instruments were developed to measure the Earth's shape, density and gravity field, as well as the components of the water cycle. In the 20th century, geophysical methods were developed for remote exploration of the solid Earth and the ocean, and geophysics played an essential role in the development of the theory of plate tectonics.

Geophysics is applied to societal needs, such as mineral resources, mitigation of natural hazards and environmental protection. Geophysical survey data are used to analyze potential petroleum reservoirs and mineral deposits, locate groundwater, find archaeological relics, determine the thickness of glaciers and soils, and assess sites for environmental remediation.

Physical Phenomena

Geophysics is a highly interdisciplinary subject, and geophysicists contribute to every area of the Earth sciences. To provide a clearer idea of what constitutes geophysics, this section describes phenomena that are studied in physics and how they relate to the Earth and its surroundings.

Gravity

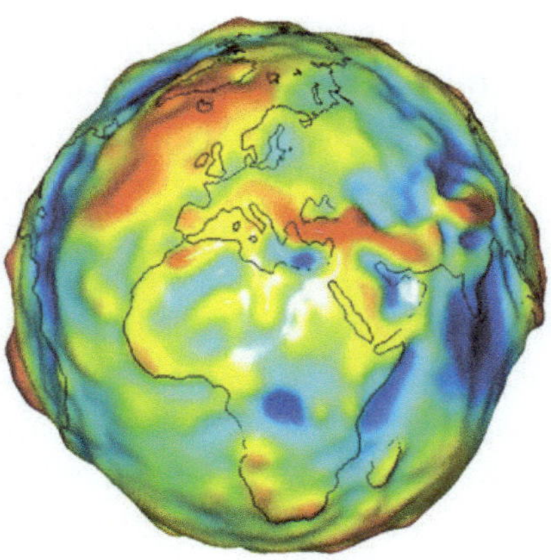

A map of deviations in gravity from a perfectly smooth, idealized Earth.

The gravitational pull of the Moon and Sun give rise to two high tides and two low tides every lunar day, or every 24 hours and 50 minutes. Therefore, there is a gap of 12 hours and 25 minutes between every high tide and between every low tide.

Gravitational forces make rocks press down on deeper rocks, increasing their density as the depth increases. Measurements of gravitational acceleration and gravitational potential at the Earth's surface and above it can be used to look for mineral deposits. The surface gravitational field provides information on the dynamics of tectonic plates. The geopotential surface called the geoid is one definition of the shape of the Earth. The geoid would be the global mean sea level if the oceans were in equilibrium and could be extended through the continents (such as with very narrow canals).

Heat Flow

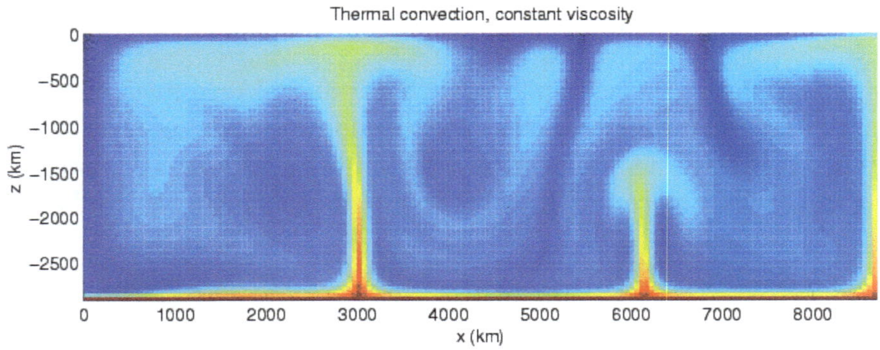

A model of thermal convection in the Earth's mantle. The thin red columns are mantle plumes.

The Earth is cooling, and the resulting heat flow generates the Earth's magnetic field through the geodynamo and plate tectonics through mantle convection. The main sources of heat are the primordial heat and radioactivity, although there are also contributions from phase transitions. Heat is mostly carried to the surface by thermal convection, although there are two thermal boundary layers – the core-mantle boundary and the lithosphere – in which heat is transported by conduction. Some heat is carried up from the bottom of the mantle by mantle plumes. The heat flow at the Earth's surface is about 4.2×10^{13} W, and it is a potential source of geothermal energy.

Vibrations

Seismic waves are vibrations that travel through the Earth's interior or along its surface. The entire Earth can also oscillate in forms that are called normal modes or free oscillations of the Earth. Ground motions from waves or normal modes are measured using seismographs. If the waves come from a localized source such as an earthquake or explosion, measurements at more than one location can be used to locate the source. The locations of earthquakes provide information on plate tectonics and mantle convection.

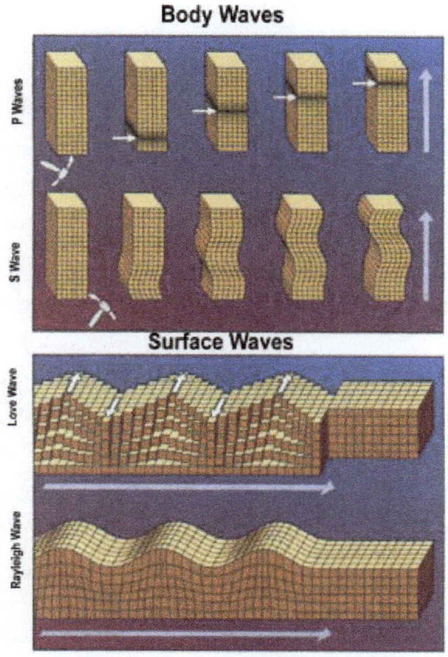

Illustration of the deformations of a block by body waves and surface waves.

Measurements of seismic waves are a source of information on the region that the waves travel through. If the density or composition of the rock changes suddenly, some waves are reflected. Reflections can provide information on near-surface structure. Changes in the travel direction, called refraction, can be used to infer the deep structure of the Earth.

Earthquakes pose a risk to humans. Understanding their mechanisms, which depend on the type of earthquake (e.g., intraplate or deep focus), can lead to better estimates of earthquake risk and improvements in earthquake engineering.

Electricity

Although we mainly notice electricity during thunderstorms, there is always a downward electric field near the surface that averages 120 V m^{-1}. Relative to the solid Earth, the atmosphere has a net positive charge due to bombardment by cosmic rays. A current of about 1800 A flows in the global circuit. It flows downward from the ionosphere over most of the Earth and back upwards through thunderstorms. The flow is manifested by lightning below the clouds and sprites above.

A variety of electric methods are used in geophysical survey. Some measure spontaneous potential, a potential that arises in the ground because of man-made or natural disturbances. Telluric currents flow in Earth and the oceans. They have two causes: electromagnetic induction by the time-varying, external-origin geomagnetic field and motion of conducting bodies (such as seawater) across the Earth's permanent magnetic field. The distribution of telluric current density can be used to detect variations in electrical resistivity of underground structures. Geophysicists can also provide the electric current themselves.

Electromagnetic Waves

Electromagnetic waves occur in the ionosphere and magnetosphere as well as the Earth's outer core. Dawn chorus is believed to be caused by high-energy electrons that get caught in the Van Allen radiation belt. Whistlers are produced by lightning strikes. Hiss may be generated by both. Electromagnetic waves may also be generated by earthquakes.

In the Earth's outer core, electric currents in the highly conductive liquid iron create magnetic fields by electromagnetic induction. Alfvén waves are magnetohydrodynamic waves in the magnetosphere or the Earth's core. In the core, they probably have little observable effect on the geomagnetic field, but slower waves such as magnetic Rossby waves may be one source of geomagnetic secular variation.

Electromagnetic methods that are used for geophysical survey include transient electromagnetics and magnetotellurics.

Magnetism

The Earth's magnetic field protects the Earth from the deadly solar wind and has long been used for navigation. It originates in the fluid motions of the Earth's outer core. The magnetic field in the upper atmosphere gives rise to the auroras.

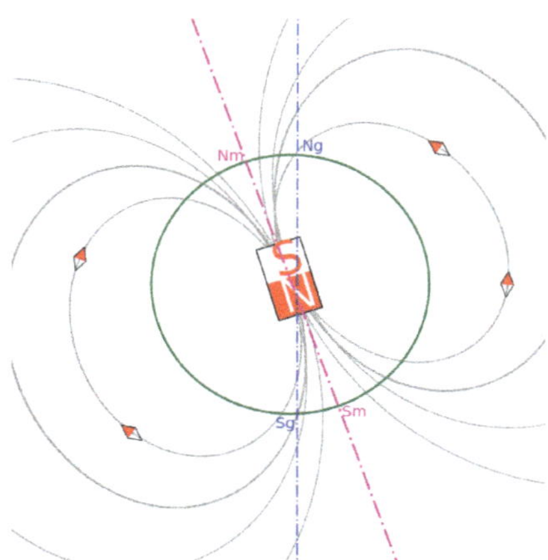

Earth's dipole axis (pink line) is tilted away from the rotational axis (blue line).

The Earth's field is roughly like a tilted dipole, but it changes over time (a phenomenon called geomagnetic secular variation). Mostly the geomagnetic pole stays near the geographic pole, but at random intervals averaging 440,000 to a million years or so, the polarity of the Earth's field reverses. These geomagnetic reversals, analyzed within a Geomagnetic Polarity Time Scale, contain 184 polarity intervals in the last 83 million years, with change in frequency over time, with the most recent brief complete reversal of the Laschamp event occurring 41,000 years ago during the last glacial period. Geologists observed geomagnetic reversal recorded in volcanic rocks, through magnetostratigraphy correlation and their signature can be seen as parallel linear magnetic anomaly stripes on the seafloor. These stripes provide quantitative information on seafloor spreading, a part of plate tectonics. They are the basis of magnetostratigraphy, which correlates magnetic reversals with other stratigraphies to construct geologic time scales. In addition, the magnetization in rocks can be used to measure the motion of continents.

Radioactivity

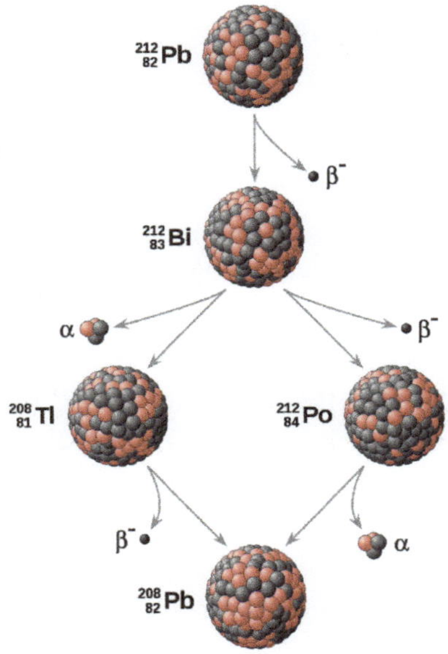

Example of a radioactive decay chain.

Radioactive decay accounts for about 80% of the Earth's internal heat, powering the geodynamo and plate tectonics. The main heat-producing isotopes are potassium-40, uranium-238, uranium-235, and thorium-232. Radioactive elements are used for radiometric dating, the primary method for establishing an absolute time scale in geochronology. Unstable isotopes decay at predictable rates, and the decay rates of different isotopes cover several orders of magnitude, so radioactive decay can be used to accurately date both recent events and events in past geologic eras. Radiometric mapping using ground and airborne gamma spectrometry can be used to map the concentration and distribution of radioisotopes near the Earth's surface, which is useful for mapping lithology and alteration.

Fluid Dynamics

Fluid motions occur in the magnetosphere, atmosphere, ocean, mantle and core. Even the mantle, though it has an enormous viscosity, flows like a fluid over long time intervals. This flow is reflected in phenomena such as isostasy, post-glacial rebound and mantle plumes. The mantle flow drives plate tectonics and the flow in the Earth's core drives the geodynamo.

Geophysical fluid dynamics is a primary tool in physical oceanography and meteorology. The rotation of the Earth has profound effects on the Earth's fluid dynamics, often due to the Coriolis effect. In the atmosphere it gives rise to large-scale patterns like Rossby waves and determines the basic circulation patterns of storms. In the ocean they drive large-scale circulation patterns as well as Kelvin waves and Ekman spirals at the ocean surface. In the Earth's core, the circulation of the molten iron is structured by Taylor columns.

Waves and other phenomena in the magnetosphere can be modeled using magnetohydrodynamics.

Mineral Physics

The physical properties of minerals must be understood to infer the composition of the Earth's interior from seismology, the geothermal gradient and other sources of information. Mineral physicists study the elastic properties of minerals; their high-pressure phase diagrams, melting points and equations of state at high pressure; and the rheological properties of rocks, or their ability to flow. Deformation of rocks by creep make flow possible, although over short times the rocks are brittle. The viscosity of rocks is affected by temperature and pressure, and in turn determines the rates at which tectonic plates move.

Water is a very complex substance and its unique properties are essential for life. Its physical properties shape the hydrosphere and are an essential part of the water cycle and climate. Its thermodynamic properties determine evaporation and the thermal gradient in the atmosphere. The many types of precipitation involve a complex mixture of processes such as coalescence, supercooling and supersaturation. Some precipitated water becomes groundwater, and groundwater flow includes phenomena such as percolation, while the conductivity of water makes electrical and electromagnetic methods useful for tracking groundwater flow. Physical properties of water such as salinity have a large effect on its motion in the oceans.

The many phases of ice form the cryosphere and come in forms like ice sheets, glaciers, sea ice, freshwater ice, snow, and frozen ground (or permafrost).

Regions of the Earth

Size and Form of the Earth

The Earth is roughly spherical, but it bulges towards the Equator, so it is roughly in the shape of an ellipsoid. This bulge is due to its rotation and is nearly consistent with an Earth in hydrostatic equilibrium. The detailed shape of the Earth, however, is also affected by the distribution of continents and ocean basins, and to some extent by the dynamics of the plates.

Structure of the Interior

Evidence from seismology, heat flow at the surface, and mineral physics is combined with the Earth's mass and moment of inertia to infer models of the Earth's interior – its composition, density, temperature, pressure. For example, the Earth's mean specific gravity (5.515) is far higher than the typical specific gravity of rocks at the surface (2.7–3.3), implying that the deeper material is denser. This is also implied by its low moment of inertia ($0.33\,M\,R^2$, compared to $0.4\,M\,R^2$ for a sphere of constant density). However, some of the density increase is compression under the enormous pressures inside the Earth. The effect of pressure can be calculated using the Adams–Williamson equation. The conclusion is that pressure alone cannot account for the increase in density. Instead, we know that the Earth's core is composed of an alloy of iron and other minerals.

Reconstructions of seismic waves in the deep interior of the Earth show that there are no S-waves in the outer core. This indicates that the outer core is liquid, because liquids cannot support shear. The outer core is liquid, and the motion of this highly conductive fluid generates the Earth's field. The inner core, however, is solid because of the enormous pressure.

Reconstruction of seismic reflections in the deep interior indicate some major discontinuities in seismic velocities that demarcate the major zones of the Earth: inner core, outer core, mantle, lithosphere and crust. The mantle itself is divided into the upper mantle, transition zone, lower mantle and D'' layer. Between the crust and the mantle is the Mohorovičić discontinuity.

The seismic model of the Earth does not by itself determine the composition of the layers. For a complete model of the Earth, mineral physics is needed to interpret seismic velocities in terms of composition. The mineral properties are temperature-dependent, so the geotherm must also be determined. This requires physical theory for thermal conduction and convection and the heat contribution of radioactive elements. The main model for the radial structure of the interior of the Earth is the preliminary reference Earth model (PREM). Some parts of this model have been updated by recent findings in mineral physics and supplemented by seismic tomography. The mantle is mainly composed of silicates, and the boundaries between layers of the mantle are consistent with phase transitions.

The mantle acts as a solid for seismic waves, but under high pressures and temperatures it deforms so that over millions of years it acts like a liquid. This makes plate tectonics possible. Geodynamics is the study of the fluid flow in the mantle and core.

Magnetosphere

If a planet's magnetic field is strong enough, its interaction with the solar wind forms a magnetosphere. Early space probes mapped out the gross dimensions of the Earth's magnetic field, which extends about 10 Earth radii towards the Sun. The solar wind, a stream of charged particles, streams out and around the terrestrial magnetic field, and continues behind the magnetic tail, hundreds of Earth radii downstream. Inside the magnetosphere, there are relatively dense regions of solar wind particles called the Van Allen radiation belts.

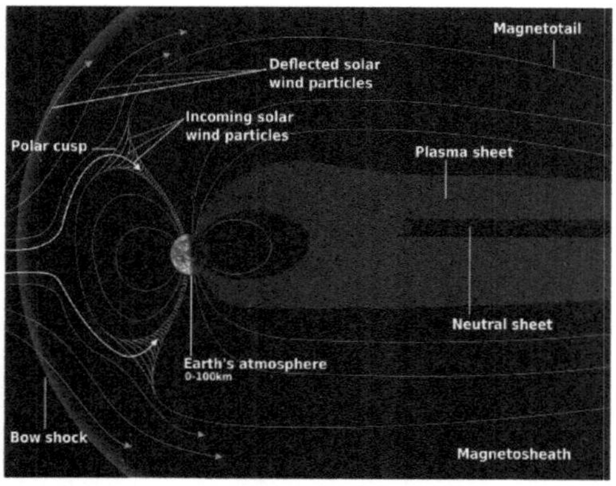

Schematic of Earth's magnetosphere. The solar wind flows from left to right.

Methods

Geodesy

Geophysical measurements are generally at a particular time and place. Accurate measurements of position, along with earth deformation and gravity, are the province of geodesy. While geodesy and geophysics are separate fields, the two are so closely connected that many scientific organizations such as the American Geophysical Union, the Canadian Geophysical Union and the International Union of Geodesy and Geophysics encompass both.

Absolute positions are most frequently determined using the global positioning system (GPS). A three-dimensional position is calculated using messages from four or more visible satellites and referred to the 1980 Geodetic Reference System. An alternative, optical astronomy, combines astronomical coordinates and the local gravity vector to get geodetic coordinates. This method only provides the position in two coordinates and is more difficult to use than GPS. However, it is useful for measuring motions of the Earth such as nutation and Chandler wobble. Relative positions of two or more points can be determined using very-long-baseline interferometry.

Gravity measurements became part of geodesy because they were needed to related measurements at the surface of the Earth to the reference coordinate system. Gravity measurements on land can be made using gravimeters deployed either on the surface or in helicopter flyovers. Since the 1960s, the Earth's gravity field has been measured by analyzing the motion of satellites. Sea level can also be measured by satellites using radar altimetry, contributing to a more accurate geoid. In 2002, NASA launched the Gravity Recovery and Climate Experiment (GRACE), wherein two twin satellites map variations in Earth's gravity field by making measurements of the distance between the two satellites using GPS and a microwave ranging system. Gravity variations detected by GRACE include those caused by changes in ocean currents; runoff and ground water depletion; melting ice sheets and glaciers.

Space Probes

Space probes made it possible to collect data from not only the visible light region, but in other areas of the electromagnetic spectrum. The planets can be characterized by their force fields: gravity and their magnetic fields, which are studied through geophysics and space physics.

Measuring the changes in acceleration experienced by spacecraft as they orbit has allowed fine details of the gravity fields of the planets to be mapped. For example, in the 1970s, the gravity field disturbances above lunar maria were measured through lunar orbiters, which led to the discovery of concentrations of mass, mascons, beneath the Imbrium, Serenitatis, Crisium, Nectaris and Humorum basins.

History

Geophysics emerged as a separate discipline only in the 19th century, from the intersection of physical geography, geology, astronomy, meteorology, and physics. However, many geophysical phenomena – such as the Earth's magnetic field and earthquakes – have been investigated since the ancient era.

Ancient and Classical Eras

Replica of Zhang Heng's seismoscope, possibly the first contribution to seismology.

The magnetic compass existed in China back as far as the fourth century BC. It was used as much for feng shui as for navigation on land. It was not until good steel needles could be forged that compasses were used for navigation at sea; before that, they could not retain their magnetism long enough to be useful. The first mention of a compass in Europe was in 1190 AD.

In circa 240 BC, Eratosthenes of Cyrene deduced that the Earth was round and measured the circumference of the Earth, using trigonometry and the angle of the Sun at more than one latitude in Egypt. He developed a system of latitude and longitude.

Perhaps the earliest contribution to seismology was the invention of a seismoscope by the prolific inventor Zhang Heng in 132 AD. This instrument was designed to drop a bronze ball from the mouth of a dragon into the mouth of a toad. By looking at which of eight toads had the ball, one could determine the direction of the earthquake. It was 1571 years before the first design for a seismoscope was published in Europe, by Jean de la Hautefeuille. It was never built.

Beginnings of Modern Science

One of the publications that marked the beginning of modern science was William Gilbert's *De Magnete* (1600), a report of a series of meticulous experiments in magnetism. Gilbert deduced that compasses point north because the Earth itself is magnetic.

In 1687 Isaac Newton published his *Principia*, which not only laid the foundations for classical mechanics and gravitation but also explained a variety of geophysical phenomena such as the tides and the precession of the equinox.

The first seismometer, an instrument capable of keeping a continuous record of seismic activity, was built by James Forbes in 1844.

Geochemistry

Geochemistry is the science that uses the tools and principles of chemistry to explain the mechanisms behind major geological systems such as the Earth's crust and its oceans. The realm of geochemistry extends beyond the Earth, encompassing the entire Solar System and has made important contributions to the understanding of a number of processes including mantle convection, the formation of planets and the origins of granite and basalt.

History

The term *geochemistry* was first used by the Swiss-German chemist Christian Friedrich Schönbein in 1838. In his paper, Schönbein predicted the birth of a new field of study, stating:

"In a word, a comparative geochemistry ought to be launched, before geochemistry can become geology, and before the mystery of the genesis of our planets and their inorganic matter may be revealed."

The field began to be realised a short time after Schönbein's work, but his term - 'geochemistry' - was initially used neither by geologists nor chemists and there was much debate over which of the two sciences should be the dominant partner. There was little collaboration between geologists and chemists and the field of geochemistry remained small and unrecognised. In the late 19th Century a Swiss man by the name of Victor Goldschmidt was born, who later became known as the father of geochemistry. His paper, Geochemische Verteilungsgesetze der Elemente, on the distribution of elements in nature, has been referred to as the start of geochemistry.

During the early 20th Century, a number of geochemists produced work that began to popularise the field, including Frank Wigglesworth Clarke who had begun to investigate the abundances of various elements within the Earth and how the quantities were related to atomic weight. The composition of meteorites and their differences to terrestrial rocks was being investigated as early as 1850 and in 1901, Oliver C. Farrington hypothesised although there were differences, that the relative abundances should still be the same. This was the beginnings of the field of cosmochemistry and has contributed much of what we know about the formation of the Earth and the Solar System.

Subfields

Some subsets of geochemistry are:

1. Isotope geochemistry involves the determination of the relative and absolute concentrations of the elements and their isotopes in the earth and on earth's surface.

2. Examination of the distribution and movements of elements in different parts of the earth (crust, mantle, hydrosphere etc.) and in minerals with the goal to determine the underlying system of distribution and movement.

3. Cosmochemistry includes the analysis of the distribution of elements and their isotopes in the cosmos.

4. Biogeochemistry is the field of study focusing on the effect of life on the chemistry of the earth.

5. Organic geochemistry involves the study of the role of processes and compounds that are derived from living or once-living organisms.

6. Aqueous geochemistry studies the role of various elements in watersheds, including copper, sulfur, mercury, and how elemental fluxes are exchanged through atmospheric-terrestrial-aquatic interactions.

7. Regional, environmental and exploration geochemistry includes applications to environmental, hydrological and mineral exploration studies.

8. Photogeochemistry is the study of light-induced chemical reactions that occur or may occur among natural components of the earth's surface.

Victor Goldschmidt is considered by most to be the father of modern geochemistry and the ideas of the subject were formed by him in a series of publications from 1922 under the title 'Geochemische Verteilungsgesetze der Elemente' (geochemical laws of distribution of the elements).

Chemical Characteristics

The more common rock constituents are nearly all oxides; chlorides, sulfides and fluorides are the only important exceptions to this and their total amount in any rock is usually much less than 1%. F. W. Clarke has calculated that a little more than 47% of the Earth's crust consists of oxygen. It occurs principally in combination as oxides, of which the chief are silica, alumina, iron oxides, and various carbonates (calcium carbonate, magnesium carbonate, sodium carbonate, and potassium carbonate). The silica functions principally as an acid, forming silicates, and all the commonest minerals of igneous rocks are of this nature. From a computation based on 1672 analyses of numerous kinds of rocks Clarke arrived at the following as the average percentage composition of the Earth's crust: SiO_2=59.71, Al_2O_3=15.41, Fe_2O_3=2.63, FeO=3.52, MgO=4.36, CaO=4.90, Na_2O=3.55, K_2O=2.80, H_2O=1.52, TiO_2=0.60, P_2O_5=0.22, (total 99.22%). All the other constituents occur only in very small quantities, usually much less than 1%.

These oxides combine in a haphazard way. For example, potash (potassium carbonate) and soda (sodium carbonate) combine to produce feldspars. In some cases they may take other forms, such as nepheline, leucite, and muscovite, but in the great majority of instances they are found as feldspar. Phosphoric acid with lime (calcium carbonate) forms apatite. Titanium dioxide with ferrous oxide gives rise to ilmenite. Part of the lime forms lime feldspar. Magnesium carbonate and iron oxides with silica crystallize as olivine or enstatite, or with alumina and lime form the complex ferro-magnesian silicates of which the pyroxenes, amphiboles, and biotites are the chief. Any excess of silica above what is required to neutralize the bases will separate out as quartz; excess of alumina crystallizes as corundum. These must be regarded only as general tendencies. It is possible, by rock analysis, to say approximately what minerals the rock contains, but there are numerous exceptions to any rule.

Mineral Constitution

Except in acid or siliceous igneous rocks containing greater than 66% of silica, known as felsic rocks, quartz is not abundant in igneous rocks. In basic rocks (containing 20%

of silica or less) it is rare for them to contain as much silicon, these are referred to as mafic rocks. If magnesium and iron are above average while silica is low, olivine may be expected; where silica is present in greater quantity over ferro-magnesian minerals, such as augite, hornblende, enstatite or biotite, occur rather than olivine. Unless potash is high and silica relatively low, leucite will not be present, for leucite does not occur with free quartz. Nepheline, likewise, is usually found in rocks with much soda and comparatively little silica. With high alkalis, soda-bearing pyroxenes and amphiboles may be present. The lower the percentage of silica and alkali's, the greater is the prevalence of plagioclase feldspar as contracted with soda or potash feldspar.

Earth's crust is composed of 90% silicate minerals and their abundance in the Earth is as follows: plagioclase feldspar (39%), alkali feldspar (12%), quartz (12%), pyroxene (11%), amphiboles (5%), micas (5%), clay minerals (5%); the remaining silicate minerals make up another 3% of Earth's crust. Only 8% of the Earth is composed of non-silicate minerals such as carbonates, oxides, and sulfides.

The other determining factor, namely the physical conditions attending consolidation, plays on the whole a smaller part, yet is by no means negligible. Certain minerals are practically confined to deep-seated intrusive rocks, e.g., microcline, muscovite, diallage. Leucite is very rare in plutonic masses; many minerals have special peculiarities in microscopic character according to whether they crystallized in depth or near the surface, e.g., hypersthene, orthoclase, quartz. There are some curious instances of rocks having the same chemical composition, but consisting of entirely different minerals, e.g., the hornblendite of Gran, in Norway, which contains only hornblende, has the same composition as some of the camptonites of the same locality that contain feldspar and hornblende of a different variety. In this connection we may repeat what has been said above about the corrosion of porphyritic minerals in igneous rocks. In rhyolites and trachytes, early crystals of hornblende and biotite may be found in great numbers partially converted into augite and magnetite. Hornblende and biotite were stable under the pressures and other conditions below the surface, but unstable at higher levels. In the ground-mass of these rocks, augite is almost universally present. But the plutonic representatives of the same magma, granite and syenite contain biotite and hornblende far more commonly than augite.

Felsic, Intermediate and Mafic Igneous Rocks

Those rocks that contain the most silica, and on crystallizing yield free quartz, form a group generally designated the "felsic" rocks. Those again that contain least silica and most magnesia and iron, so that quartz is absent while olivine is usually abundant, form the "mafic" group. The "intermediate" rocks include those characterized by the general absence of both quartz and olivine. An important subdivision of these contains a very high percentage of alkalis, especially soda, and consequently has minerals such as nepheline and leucite not common in other rocks. It is often separated from the others as the "alkali" or "soda" rocks, and there is a corresponding series of mafic rocks.

Lastly a small sub-group rich in olivine and without feldspar has been called the "ultramafic" rocks. They have very low percentages of silica but much iron and magnesia.

Except these last, practically all rocks contain felspars or feldspathoid minerals. In the acid rocks the common feldspars are orthoclase, perthite, microcline, and oligoclase—all having much silica and alkalis. In the mafic rocks labradorite, anorthite and bytownite prevail, being rich in lime and poor in silica, potash and soda. Augite is the most common ferro-magnesian in mafic rocks, but biotite and hornblende are on the whole more frequent in felsic rocks.

	Acid	**Intermediate**		**Mafic**	**Ultramafic**
Most Common Minerals	**Quartz Orthoclase (and Oligoclase), Mica, Hornblende, Augite**	**Little or no Quartz: Orthoclase hornblende, Augite, Biotite**	**Little or no Quartz: Plagioclase Hornblende, Augite, Biotite**	**No Quartz Plagioclase Augite, Olivine**	**No Felspar Augite, Hornblende, Olivine**
Plutonic or Abyssal type	Granite	Syenite	Diorite	Gabbro	Peridotite
Intrusive or Hypabyssal type	Quartz-porphyry	Orthoclase-porphyry	Porphyrite	Dolerite	Picrite
Lavas or Effusive type	Rhyolite, Obsidian	Trachyte	Andesite	Basalt	Limburgite

Rocks that contain leucite or nepheline, either partly or a wholly replacing felspar, are not included in this table. They are essentially of intermediate or of mafic character. We might in consequence regard them as varieties of syenite, diorite, gabbro, etc., in which feldspathoid minerals occur, and indeed there are many transitions between syenites of ordinary type and nepheline — or leucite — syenite, and between gabbro or dolerite and theralite or essexite. But, as many minerals develop in these "alkali" rocks that are uncommon elsewhere, it is convenient in a purely formal classification like that outlined here to treat the whole assemblage as a distinct series.

This classification is based essentially on the mineralogical constitution of the igneous rocks. Any chemical distinctions between the different groups, though implied, are relegated to a subordinate position. It is admittedly artificial but it has grown up with the growth of the science and is still adopted as the basis on which more minute subdivisions are erected. The subdivisions are by no means of equal value. The syenites, for example, and the peridotites, are far less important than the granites, diorites and gabbros. Moreover, the effusive andesites do not always correspond to the plutonic diorites but partly also to the gabbros. As the different kinds of rock, regarded as aggregates of minerals, pass gradually into one another, transitional types are very common and are often so important as to receive special names. The quartz-syenites and nordmarkites may be interposed between granite and syenite, the tonalites and adamellites between granite and diorite, the monzoaites between syenite and diorite, norites and hyperites between diorite and gabbro, and so on.

Geochemistry of Trace Metals in the Ocean

Trace metals readily form complexes with major ions in the ocean, including hydroxide, carbonate, and chloride and their chemical speciation changes depending on whether the environment is oxidized or reduced. Benjamin (2002) defines complexes of metals with more than one type of ligand, other than water, as mixed-ligand-complexes. In some cases, a ligand contains more than one *donor* atom, forming very strong complexes, also called chelates (the ligand is the chelator). One of the most common chelators is EDTA (ethylenediaminetetraacetic acid), which can replace six molecules of water and form strong bonds with metals that have a plus two charge. With stronger complexation, lower activity of the free metal ion is observed. One consequence of the lower reactivity of complexed metals compared to the same concentration of free metal is that the chelation tends to stabilize metals in the aqueous solution instead of in solids.

Concentrations of the trace metals cadmium, copper, molybdenum, manganese, rhenium, uranium and vanadium in sediments record the redox history of the oceans. Within aquatic environments, cadmium(II) can either be in the form $CdCl^{+}_{(aq)}$ in oxic waters or CdS(s) in a reduced environment. Thus higher concentrations of Cd in marine sediments may indicate low redox potential conditions in the past. For copper(II), a prevalent form is $CuCl^{+}(aq)$ within oxic environments and CuS(s) and Cu_2S within reduced environments. The reduced seawater environment leads to two possible oxidation states of copper, Cu(I) and Cu(II). Molybdenum is present as the Mo(VI) oxidation state as $MoO_4{}^{2-}{}_{(aq)}$ in oxic environments. Mo(V) and Mo(IV) are present in reduced environments in the forms $MoO_2{}^{+}{}_{(aq)}$ and $MoS_{2(s)}$. Rhenium is present as the Re(VII) oxidation state as $ReO_4{}^{-}$ within oxic conditions, but is reduced to Re(IV) which may form ReO_2 or ReS_2. Uranium is in oxidation state VI in $UO_2(CO_3)_3{}^{4-}(aq)$ and is found in the reduced form $UO_2(s)$. Vanadium is in several forms in oxidation state V(V); $HVO_4{}^{2-}$ and $H_2VO_4{}^{-}$. Its reduced forms can include $VO_2{}^{+}$, $VO(OH)_3{}^{-}$, and $V(OH)_3$. These relative dominance of these species depends on pH.

In the water column of the ocean or deep lakes, vertical profiles of dissolved trace metals are characterized as following *conservative–type*, *nutrient–type*, or *scavenged–type* distributions. Across these three distributions, trace metals have different residence times and are used to varying extents by planktonic microorganisms. Trace metals with conservative-type distributions have high concentrations relative to their biological use. One example of a trace metal with a conservative-type distribution is molybdenum. It has a residence time within the oceans of around 8×10^5 years and is generally present as the molybdate anion (MoO_4^{2-}). Molybdenum interacts weakly with particles and displays an almost uniform vertical profile in the ocean. Relative to the abundance of molybdenum in the ocean, the amount required as a metal cofactor for enzymes in marine phytoplankton is negligible.

Trace metals with nutrient-type distributions are strongly associated with the internal cycles of particulate organic matter, especially the assimilation by plankton. The lowest dissolved concentrations of these metals are at the surface of the ocean, where they are assimilated by plankton. As dissolution and decomposition occur at greater depths, concentrations of these trace metals increase. Residence times of these metals, such as zinc, are several thousand to one hundred thousand years. Finally, an example of a scavenged-type trace metal is aluminium, which has strong interactions with particles as well as a short residence time in the ocean. The residence times of scavenged-type trace metals are around 100 to 1000 years. The concentrations of these metals are highest around bottom sediments, hydrothermal vents, and rivers. For aluminium, atmospheric dust provides the greatest source of external inputs into the ocean.

Iron and copper show hybrid distributions in the ocean. They are influenced by recycling and intense scavenging. Iron is a limiting nutrient in vast areas of the oceans, and is found in high abundance along with manganese near hydrothermal vents. Here, many iron precipitates are found, mostly in the forms of iron sulfides and oxidized iron oxyhydroxide compounds. Concentrations of iron near hydrothermal vents can be up to one million times the concentrations found in the open ocean.

Using electrochemical techniques, it is possible to show that bioactive trace metals (zinc, cobalt, cadmium, iron and copper) are bound by organic ligands in surface seawater. These ligand complexes serve to lower the bioavailability of trace metals within the ocean. For example, copper, which may be toxic to open ocean phytoplankton and bacteria, can form organic complexes. The formation of these complexes reduces the concentrations of bioavailable inorganic complexes of copper that could be toxic to sea life at high concentrations. Unlike copper, zinc toxicity in marine phytoplankton is low and there is no advantage to increasing the organic binding of Zn^{2+}. In high nutrient-low chlorophyll regions, iron is the limiting nutrient, with the dominant species being strong organic complexes of Fe(III).

References

- Burger, H. Robert; Sheehan, Anne F.; Jones, Craig H. (2006). Introduction to applied geophysics : exploring the shallow subsurface. New York: W.W. Norton. ISBN 0-393-92637-0.
- Krumbein, Wolfgang E., ed. (1978). Environmental biogeochemistry and geomicrobiology. Ann Arbor, Mich.: Ann Arbor Science Publ. ISBN 0-250-40218-1.
- Hubbard, Bryn; Glasser, Neil (2005). Field techniques in glaciology and glacial geomorphology. Chichester, England: J. Wiley. ISBN 0-470-84426-4.
- Shepherd, T.J.; Rankin, A.H.; Alderton, D.H.M. (1985). A practical guide to fluid inclusion stud-

ies. Glasgow: Blackie. ISBN 0-412-00601-4.

- Spear, Frank S. (1995). Metamorphic phase equilibria and pressure-temperature-time paths. Washington, DC: Mineralogical Soc. of America. ISBN 978-0-939950-34-8.
- Vai, Gian Battista; Cavazza, William (2003). Four centuries of the word geology: Ulisse Aldrovandi 1603 in Bologna. Minerva. ISBN 978-88-7381-056-8.
- Eratosthenes (2010). Eratosthenes' "Geography". Fragments collected and translated, with commentary and additional material by Duane W. Roller. Princeton University Press. ISBN 978-0-691-14267-8.
- Sheriff, Robert E. (1991). "Geophysics". Encyclopedic Dictionary of Exploration Geophysics (3rd ed.). Society of Exploration. ISBN 978-1-56080-018-7.
- Stein, Seth; Wysession, Michael (2003). An introduction to seismology, earthquakes, and earth structure. Wiley-Blackwell. ISBN 0-86542-078-5.
- Telford, William Murray; Geldart, L. P.; Sheriff, Robert E. (1990). Applied geophysics. Cambridge University Press. ISBN 978-0-521-33938-4.
- Poirier, Jean-Paul (2000). Introduction to the Physics of the Earth's Interior. Cambridge Topics in Mineral Physics & Chemistry. Cambridge University Press. ISBN 0-521-66313-X.
- Sadava, David; Heller, H. Craig; Hillis, David M.; Berenbaum, May (2009). Life: The Science of Biology. Macmillan. ISBN 978-1-4292-1962-4.

8

Plate Tectonics: A Significant Study

The theory that the Earth's outer shield is divided into several plates, some of these plates being large in size while some being small, is identified as plate tectonics. In plate tectonics, divergent boundary is a linear feature that exists between two tectonic plates, which are moving away from each other. This chapter will provide an integrated understanding of plate tectonics.

Plate Tectonics

Plate tectonics (from the Late Latin *tectonicus*, "pertaining to building") is a scientific theory describing the large-scale motion of 7 large plates and the movements of a larger number of smaller plates of the Earth's lithosphere, over the last 100's of millions of years. The theoretical model builds on the concept of continental drift developed during the first few decades of the 20th century. The geoscientific community accepted plate-tectonic theory after seafloor spreading was validated in the late 1950s and early 1960s.

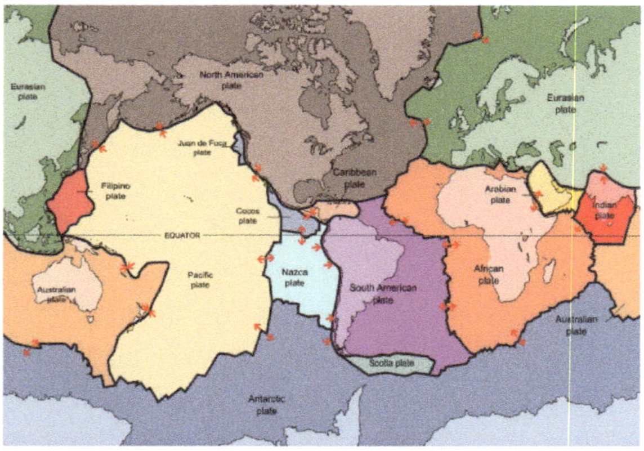

The tectonic plates of the world were mapped in the second half of the 20th century.

The lithosphere, which is the rigid outermost shell of a planet (the crust and upper mantle), is broken up into tectonic plates. The Earth's lithosphere is composed of seven or eight major plates (depending on how they are defined) and many minor plates. Where the plates meet, their relative motion determines the type of boundary: convergent, divergent, or transform. Earthquakes, volcanic activity, mountain-building, and oceanic trench formation occur along these plate boundaries. The relative movement of the plates typically ranges from zero to 100 mm annually.

Tectonic plates are composed of oceanic lithosphere and thicker continental lithosphere, each topped by its own kind of crust. Along convergent boundaries, subduction carries plates into the mantle; the material lost is roughly balanced by the formation of new (oceanic) crust along divergent margins by seafloor spreading. In this way, the total surface of the lithosphere remains the same. This prediction of plate tectonics is also referred to as the conveyor belt principle. Earlier theories (that still have some supporters) propose gradual shrinking (contraction) or gradual expansion of the globe.

Tectonic plates are able to move because the Earth's lithosphere has greater strength than the underlying asthenosphere. Lateral density variations in the mantle result in convection. Plate movement is thought to be driven by a combination of the motion of the seafloor away from the spreading ridge (due to variations in topography and density of the crust, which result in differences in gravitational forces) and drag, with downward suction, at the subduction zones. Another explanation lies in the different forces generated by tidal forces of the Sun and Moon. The relative importance of each of these factors and their relationship to each other is unclear, and still the subject of much debate.

Key Principles

The outer layers of the Earth are divided into the lithosphere and asthenosphere. This is based on differences in mechanical properties and in the method for the transfer of heat. Mechanically, the lithosphere is cooler and more rigid, while the asthenosphere is hotter and flows more easily. In terms of heat transfer, the lithosphere loses heat by conduction, whereas the asthenosphere also transfers heat by convection and has a nearly adiabatic temperature gradient. This division should not be confused with the *chemical* subdivision of these same layers into the mantle (comprising both the asthenosphere and the mantle portion of the lithosphere) and the crust: a given piece of mantle may be part of the lithosphere or the asthenosphere at different times depending on its temperature and pressure.

The key principle of plate tectonics is that the lithosphere exists as separate and distinct *tectonic plates*, which ride on the fluid-like (visco-elastic solid) asthenosphere. Plate motions range up to a typical 10–40 mm/year (Mid-Atlantic Ridge; about as fast as fingernails grow), to about 160 mm/year (Nazca Plate; about as fast as hair grows). The driving mechanism behind this movement is described below.

Tectonic lithosphere plates consist of lithospheric mantle overlain by either or both of two types of crustal material: oceanic crust (in older texts called *sima* from silicon and magnesium) and continental crust (*sial* from silicon and aluminium). Average oceanic lithosphere is typically 100 km (62 mi) thick; its thickness is a function of its age: as time passes, it conductively cools and subjacent cooling mantle is added to its base. Because it is formed at mid-ocean ridges and spreads outwards, its thickness is therefore a function of its distance from the mid-ocean ridge where it was formed. For a typical distance that oceanic lithosphere must travel before being subducted, the thickness varies from about

6 km (4 mi) thick at mid-ocean ridges to greater than 100 km (62 mi) at subduction zones; for shorter or longer distances, the subduction zone (and therefore also the mean) thickness becomes smaller or larger, respectively. Continental lithosphere is typically ~200 km thick, though this varies considerably between basins, mountain ranges, and stable cratonic interiors of continents. The two types of crust also differ in thickness, with continental crust being considerably thicker than oceanic (35 km vs. 6 km).

The location where two plates meet is called a *plate boundary*. Plate boundaries are commonly associated with geological events such as earthquakes and the creation of topographic features such as mountains, volcanoes, mid-ocean ridges, and oceanic trenches. The majority of the world's active volcanoes occur along plate boundaries, with the Pacific Plate's Ring of Fire being the most active and widely known today. These boundaries are discussed in further detail below. Some volcanoes occur in the interiors of plates, and these have been variously attributed to internal plate deformation and to mantle plumes.

As explained above, tectonic plates may include continental crust or oceanic crust, and most plates contain both. For example, the African Plate includes the continent and parts of the floor of the Atlantic and Indian Oceans. The distinction between oceanic crust and continental crust is based on their modes of formation. Oceanic crust is formed at sea-floor spreading centers, and continental crust is formed through arc volcanism and accretion of terranes through tectonic processes, though some of these terranes may contain ophiolite sequences, which are pieces of oceanic crust considered to be part of the continent when they exit the standard cycle of formation and spreading centers and subduction beneath continents. Oceanic crust is also denser than continental crust owing to their different compositions. Oceanic crust is denser because it has less silicon and more heavier elements ("mafic") than continental crust ("felsic"). As a result of this density stratification, oceanic crust generally lies below sea level (for example most of the Pacific Plate), while continental crust buoyantly projects above sea level.

Types of Plate Boundaries

Three types of plate boundaries exist, with a fourth, mixed type, characterized by the way the plates move relative to each other. They are associated with different types of surface phenomena. The different types of plate boundaries are:

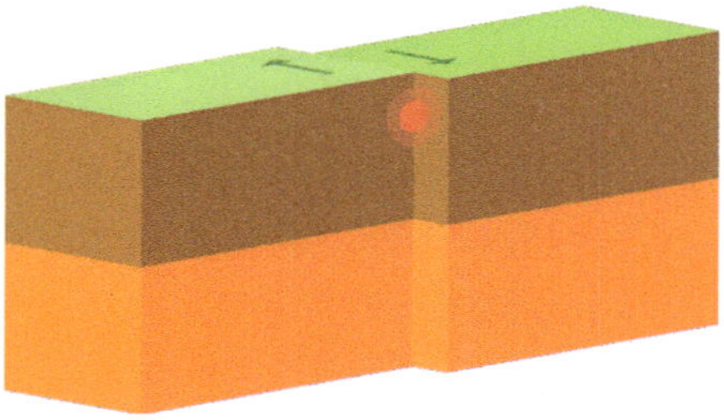

Transform boundary

1. *Transform boundaries (Conservative)* occur where two lithospheric plates slide, or perhaps more accurately, grind past each other along transform faults, where plates are neither created nor destroyed. The relative motion of the two plates is either sinistral (left side toward the observer) or dextral (right side toward the observer). Transform faults occur across a spreading center. Strong earthquakes can occur along a fault. The San Andreas Fault in California is an example of a transform boundary exhibiting dextral motion.

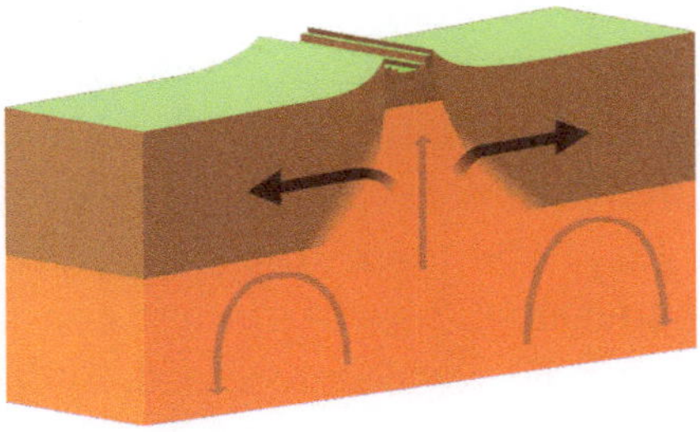

Divergent boundary

1. *Divergent boundaries (Constructive)* occur where two plates slide apart from each other. At zones of ocean-to-ocean rifting, divergent boundaries form by seafloor spreading, allowing for the formation of new ocean basin. As the ocean plate splits, the ridge forms at the spreading center, the ocean basin expands, and finally, the plate area increases causing many small volcanoes and/or shallow earthquakes. At zones of continent-to-continent rifting, divergent boundaries may cause new ocean basin to form as the continent splits, spreads, the central rift collapses, and ocean fills the basin. Active zones of Mid-ocean ridges (e.g., Mid-Atlantic Ridge and East Pacific Rise), and continent-to-continent rifting (such as Africa's East African Rift and Valley, Red Sea) are examples of divergent boundaries.

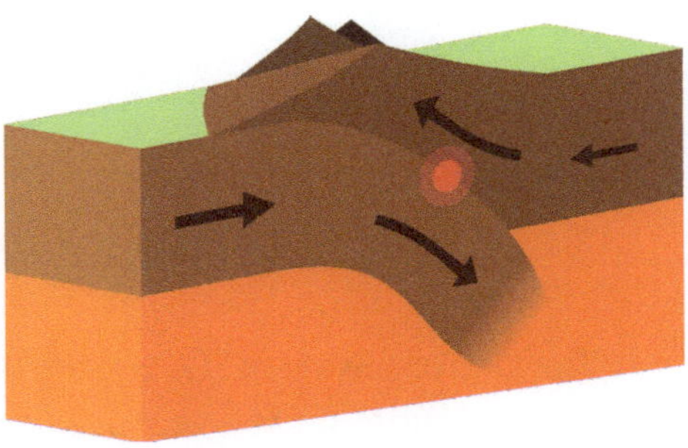

Convergent boundary

1. *Convergent boundaries (Destructive)* (or *active margins*) occur where two plates slide toward each other to form either a subduction zone (one plate moving underneath the other) or a continental collision. At zones of ocean-to-continent subduction (e.g. the Andes mountain range in South America, and the Cascade Mountains in Western United States), the dense oceanic lithosphere plunges beneath the less dense continent. Earthquakes trace the path of the down-

ward-moving plate as it descends into asthenosphere, a trench forms, and as the subducted plate is heated it releases volatiles, mostly water from hydrous minerals, into the surrounding mantle. The addition of water lowers the melting point of the mantle material above the subducting slab, causing it to melt. The magma that results typically leads to volcanism. At zones of ocean-to-ocean subduction (e.g. Aleutian islands, Mariana Islands, and the Japanese island arc), older, cooler, denser crust slips beneath less dense crust. This causes earthquakes and a deep trench to form in an arc shape. The upper mantle of the subducted plate then heats and magma rises to form curving chains of volcanic islands. Deep marine trenches are typically associated with subduction zones, and the basins that develop along the active boundary are often called "foreland basins". Closure of ocean basins can occur at continent-to-continent boundaries (e.g., Himalayas and Alps): collision between masses of granitic continental lithosphere; neither mass is subducted; plate edges are compressed, folded, uplifted.

2. *Plate boundary zones* occur where the effects of the interactions are unclear, and the boundaries, usually occurring along a broad belt, are not well defined and may show various types of movements in different episodes.

Driving Forces of Plate Motion

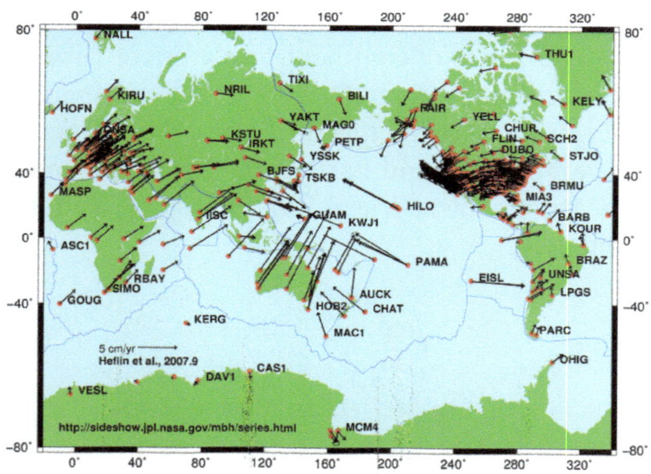

Plate motion based on Global Positioning System (GPS) satellite data from NASA JPL. The vectors show direction and magnitude of motion.

It has generally been accepted that tectonic plates are able to move because of the relative density of oceanic lithosphere and the relative weakness of the asthenosphere. Dissipation of heat from the mantle is acknowledged to be the original source of the energy required to drive plate tectonics through convection or large scale upwelling and doming. The current view, though still a matter of some debate, asserts that as a consequence, a powerful source of plate motion is generated due to the excess density of the oceanic lithosphere sinking in subduction zones. When the new crust forms at mid-ocean ridges, this oceanic lithosphere is initially less dense than the underlying asthenosphere, but it becomes denser with age as it conductively cools and thickens. The greater density of old lithosphere relative to the underlying asthenosphere allows it to sink into the deep mantle at subduction zones, providing most of the driving force for plate movement. The weakness of the asthenosphere allows the tectonic plates to move easily towards a subduction zone. Although subduction is thought to be the strongest force driving plate motions, it cannot be the only force since there are plates such as the North American Plate which are moving, yet are nowhere being subducted. The same is true for the enormous Eurasian Plate. The sources of plate motion are a matter of intensive research and discussion among scientists. One of the main points is that the kinematic pattern of the movement itself should be separated clearly from the possible geodynamic mechanism that is invoked as the driving force of the observed movement, as some patterns may be explained by more than one mechanism. In short, the driving forces advocated at the moment can be divided into three categories based on the relationship to the movement: mantle dynamics related, gravity related (mostly secondary forces).

Driving Forces Related to Mantle Dynamics

For much of the last quarter century, the leading theory of the driving force behind tectonic plate motions envisaged large scale convection currents in the upper mantle, which can be transmitted through the asthenosphere. This theory was launched by Arthur Holmes and some forerunners in the 1930s and was immediately recognized as the solution for the acceptance of the theory as originally discussed in the papers of Alfred Wegener in the early years of the century. However, despite its acceptance, it was long debated in the scientific community because the leading ("fixist") theory still envisaged a static Earth without moving continents up until the major breakthroughs of the early sixties.

Two- and three-dimensional imaging of Earth's interior (seismic tomography) shows a varying lateral density distribution throughout the mantle. Such density variations can be material (from rock chemistry), mineral (from variations in mineral structures), or thermal (through thermal expansion and contraction from heat energy). The manifestation of this varying lateral density is mantle convection from buoyancy forces.

How mantle convection directly and indirectly relates to plate motion is a matter of ongoing study and discussion in geodynamics. Somehow, this energy must be transferred to the lithosphere for tectonic plates to move. There are essentially two main types of forces that are thought to influence plate motion: friction and gravity.

- Basal drag (friction): Plate motion driven by friction between the convection currents in the asthenosphere and the more rigid overlying lithosphere.
- Slab suction (gravity): Plate motion driven by local convection currents that exert a downward pull on plates in subduction zones at ocean trenches. Slab

suction may occur in a geodynamic setting where basal tractions continue to act on the plate as it dives into the mantle (although perhaps to a greater extent acting on both the under and upper side of the slab).

Lately, the convection theory has been much debated as modern techniques based on 3D seismic tomography still fail to recognize these predicted large scale convection cells. Therefore, alternative views have been proposed:

In the theory of plume tectonics developed during the 1990s, a modified concept of mantle convection currents is used. It asserts that super plumes rise from the deeper mantle and are the drivers or substitutes of the major convection cells. These ideas, which find their roots in the early 1930s with the so-called "fixistic" ideas of the European and Russian Earth Science Schools, find resonance in the modern theories which envisage hot spots/mantle plumes which remain fixed and are overridden by oceanic and continental lithosphere plates over time and leave their traces in the geological record (though these phenomena are not invoked as real driving mechanisms, but rather as modulators). Modern theories that continue building on the older mantle doming concepts and see plate movements as a secondary phenomena are beyond the scope of this page and are discussed elsewhere (for example on the plume tectonics page).

Another theory is that the mantle flows neither in cells nor large plumes but rather as a series of channels just below the Earth's crust, which then provide basal friction to the lithosphere. This theory, called "surge tectonics", became quite popular in geophysics and geodynamics during the 1980s and 1990s. Recent research, based on three-dimensional computer modeling, suggests that plate geometry is governed by a feedback between mantle convection patterns and the strength of the lithosphere.

Driving Forces Related to Gravity

Forces related to gravity are usually invoked as secondary phenomena within the framework of a more general driving mechanism such as the various forms of mantle dynamics described above.

Gravitational sliding away from a spreading ridge: According to many authors, plate motion is driven by the higher elevation of plates at ocean ridges. As oceanic lithosphere is formed at spreading ridges from hot mantle material, it gradually cools and thickens with age (and thus adds distance from the ridge). Cool oceanic lithosphere is significantly denser than the hot mantle material from which it is derived and so with increasing thickness it gradually subsides into the mantle to compensate the greater load. The result is a slight lateral incline with increased distance from the ridge axis.

This force is regarded as a secondary force and is often referred to as "ridge push". This is a misnomer as nothing is "pushing" horizontally and tensional features are dominant along ridges. It is more accurate to refer to this mechanism as gravitational sliding as variable topography across the totality of the plate can vary considerably and the topography of spreading ridges is only the most prominent feature. Other mechanisms generating this gravitational secondary force include flexural bulging of the lithosphere before it dives underneath an adjacent plate which produces a clear topographical feature that can offset, or at least affect, the influence of topographical ocean ridges, and mantle plumes and hot spots, which are postulated to impinge on the underside of tectonic plates.

Slab-pull: Current scientific opinion is that the asthenosphere is insufficiently competent or rigid to directly cause motion by friction along the base of the lithosphere. Slab pull is therefore most widely thought to be the greatest force acting on the plates. In this current understanding, plate motion is mostly driven by the weight of cold, dense plates sinking into the mantle at trenches. Recent models indicate that trench suction plays an important role as well. However, as the North American Plate is nowhere being subducted, yet it is in motion presents a problem. The same holds for the African, Eurasian, and Antarctic plates.

Gravitational sliding away from mantle doming: According to older theories, one of the driving mechanisms of the plates is the existence of large scale asthenosphere/mantle domes which cause the gravitational sliding of lithosphere plates away from them. This gravitational sliding represents a secondary phenomenon of this basically vertically oriented mechanism. This can act on various scales, from the small scale of one island arc up to the larger scale of an entire ocean basin.

Driving forces Related to Earth rotation

Alfred Wegener, being a meteorologist, had proposed tidal forces and pole flight force as the main driving mechanisms behind continental drift; however, these forces were considered far too small to cause continental motion as the concept then was of continents plowing through oceanic crust. Therefore, Wegener later changed his position and asserted that convection currents are the main driving force of plate tectonics in the last edition of his book in 1929.

However, in the plate tectonics context (accepted since the seafloor spreading proposals of Heezen, Hess, Dietz, Morley, Vine, and Matthews during the early 1960s), the oceanic crust is suggested to be in motion *with* the continents which caused the proposals related to Earth rotation to be reconsidered. In more recent literature, these driving forces are:

1. Tidal drag due to the gravitational force the Moon (and the Sun) exerts on the crust of the Earth
2. Global deformation of the geoid due to small displacements of rotational pole with respect to the Earth's crust;

3. Other smaller deformation effects of the crust due to wobbles and spin movements of the Earth rotation on a smaller time scale.

Forces that are small and generally negligible are:

1. The Coriolis force

2. The centrifugal force, which is treated as a slight modification of gravity

For these mechanisms to be overall valid, systematic relationships should exist all over the globe between the orientation and kinematics of deformation and the geographical latitudinal and longitudinal grid of the Earth itself. Ironically, these systematic relations studies in the second half of the nineteenth century and the first half of the twentieth century underline exactly the opposite: that the plates had not moved in time, that the deformation grid was fixed with respect to the Earth equator and axis, and that gravitational driving forces were generally acting vertically and caused only local horizontal movements (the so-called pre-plate tectonic, "fixist theories"). Later studies (discussed below on this page), therefore, invoked many of the relationships recognized during this pre-plate tectonics period to support their theories.

Of the many forces discussed in this paragraph, tidal force is still highly debated and defended as a possible principle driving force of plate tectonics. The other forces are only used in global geodynamic models not using plate tectonics concepts (therefore beyond the discussions treated in this section) or proposed as minor modulations within the overall plate tectonics model.

In 1973, George W. Moore of the USGS and R. C. Bostrom presented evidence for a general westward drift of the Earth's lithosphere with respect to the mantle. He concluded that tidal forces (the tidal lag or "friction") caused by the Earth's rotation and the forces acting upon it by the Moon are a driving force for plate tectonics. As the Earth spins eastward beneath the moon, the moon's gravity ever so slightly pulls the Earth's surface layer back westward, just as proposed by Alfred Wegener (see above). In a more recent 2006 study, scientists reviewed and advocated these earlier proposed ideas. It has also been suggested recently in Lovett (2006) that this observation may also explain why Venus and Mars have no plate tectonics, as Venus has no moon and Mars' moons are too small to have significant tidal effects on the planet. In a recent paper, it was suggested that, on the other hand, it can easily be observed that many plates are moving north and eastward, and that the dominantly westward motion of the Pacific ocean basins derives simply from the eastward bias of the Pacific spreading center (which is not a predicted manifestation of such lunar forces). In the same paper the authors admit, however, that relative to the lower mantle, there is a slight westward component in the motions of all the plates. They demonstrated though that the westward drift, seen only for the past 30 Ma, is attributed to the increased dominance of the steadily growing and accelerating Pacific plate. The debate is still open.

Relative Significance of Each Driving Force Mechanism

The vector of a plate's motion is a function of all the forces acting on the plate; however, therein lies the problem regarding the degree to which each process contributes to the overall motion of each tectonic plate.

The diversity of geodynamic settings and the properties of each plate result from the impact of the various processes actively driving each individual plate. One method of dealing with this problem is to consider the relative rate at which each plate is moving as well as the evidence related to the significance of each process to the overall driving force on the plate.

One of the most significant correlations discovered to date is that lithospheric plates attached to downgoing (subducting) plates move much faster than plates not attached to subducting plates. The Pacific plate, for instance, is essentially surrounded by zones of subduction (the so-called Ring of Fire) and moves much faster than the plates of the Atlantic basin, which are attached (perhaps one could say 'welded') to adjacent continents instead of subducting plates. It is thus thought that forces associated with the downgoing plate (slab pull and slab suction) are the driving forces which determine the motion of plates, except for those plates which are not being subducted. The driving forces of plate motion continue to be active subjects of on-going research within geophysics and tectonophysics.

Development of the Theory

Summary

Detailed map showing the tectonic plates with their movement vectors.

In line with other previous and contemporaneous proposals, in 1912 the meteorologist Alfred Wegener amply described what he called continental drift, expanded in his 1915 book *The Origin of Continents and Oceans* and the scientific debate started that would end up fifty years later in the theory of plate tectonics. Starting from the idea (also expressed by his forerunners) that the present continents once formed a single land mass (which was called Pangea later on) that drifted apart, thus releasing the continents from the Earth's mantle and likening them to "icebergs" of low density granite floating on a sea of denser basalt. Supporting evidence for the idea came from the dove-tailing outlines of South America's east coast and Africa's west coast, and from the matching of the rock formations along these edges. Confirmation of their previous contiguous nature also came from the fossil plants *Glossopteris* and *Gangamopteris*, and the therapsid or mammal-like reptile *Lystrosaurus*, all widely distributed over South America, Africa, Antarctica, India and Australia. The evidence for such an erstwhile joining of these continents was patent to field geologists working in the southern hemisphere. The South African Alex du Toit put together a mass of such information in his 1937 publication *Our Wandering Continents*, and went further than Wegener in recognising the strong links between the Gondwana fragments.

But without detailed evidence and a force sufficient to drive the movement, the theory was not generally accepted: the Earth might have a solid crust and mantle and a liquid core, but there seemed to be no way that portions of the crust could move around. Distinguished scientists, such as Harold Jeffreys and Charles Schuchert, were outspoken critics of continental drift.

Despite much opposition, the view of continental drift gained support and a lively debate started between "drifters" or "mobilists" (proponents of the theory) and "fixists" (opponents). During the 1920s, 1930s and 1940s, the former reached important milestones proposing that convection currents might have driven the plate movements, and that spreading may have occurred below the sea within the oceanic crust. Concepts close to the elements now incorporated in plate tectonics were proposed by geophysicists and geologists (both fixists and mobilists) like Vening-Meinesz, Holmes, and Umbgrove.

One of the first pieces of geophysical evidence that was used to support the movement of lithospheric plates came from paleomagnetism. This is based on the fact that rocks of different ages show a variable magnetic field direction, evidenced by studies since the mid–nineteenth century. The magnetic north and south poles reverse through time, and, especially important in paleotectonic studies, the relative position of the magnetic north pole varies through time. Initially, during the first half of the twentieth century, the latter phenomenon was explained by introducing what was called "polar wander", i.e., it was assumed that the north pole location had been shifting through time. An alternative explanation, though, was that the continents had moved (shifted and rotated) relative to the north pole, and each continent, in fact, shows its own "polar wander path". During the late 1950s it was successfully shown on two occasions that these data could show the validity of continental drift: by Keith Runcorn in a paper in 1956, and by Warren Carey in a symposium held in March 1956.

The second piece of evidence in support of continental drift came during the late 1950s and early 60s from data on the bathymetry of the deep ocean floors and the nature of the oceanic crust such as magnetic properties and, more generally, with the development of marine geology which gave evidence for the association of seafloor spreading along the mid-oceanic ridges and magnetic field reversals, published between 1959 and 1963 by Heezen, Dietz, Hess, Mason, Vine & Matthews, and Morley.

Simultaneous advances in early seismic imaging techniques in and around Wadati-Benioff zones along the trenches bounding many continental margins, together with many other geophysical (e.g. gravimetric) and geological observations, showed how the oceanic crust could disappear into the mantle, providing the mechanism to balance the extension of the ocean basins with shortening along its margins.

All this evidence, both from the ocean floor and from the continental margins, made it clear around 1965 that continental drift was feasible and the theory of plate tectonics, which was defined in a series of papers between 1965 and 1967, was born, with all its extraordinary explanatory and predictive power. The theory revolutionized the Earth sciences, explaining a diverse range of geological phenomena and their implications in other studies such as paleogeography and paleobiology.

Continental Drift

In the late 19th and early 20th centuries, geologists assumed that the Earth's major features were fixed, and that most geologic features such as basin development and mountain ranges could be explained by vertical crustal movement, described in what is called the geosynclinal theory. Generally, this was placed in the context of a contracting planet Earth due to heat loss in the course of a relatively short geological time.

Alfred Wegener in Greenland in the winter of 1912-13.

It was observed as early as 1596 that the opposite coasts of the Atlantic Ocean—or, more precisely, the edges of the continental shelves—have similar shapes and seem to have once fitted together.

Since that time many theories were proposed to explain this apparent complementarity, but the assumption of a solid Earth made these various proposals difficult to accept.

The discovery of radioactivity and its associated heating properties in 1895 prompted a re-examination of the apparent age of the Earth. This had previously been estimated by its cooling rate and assumption the Earth's surface radiated like a black body. Those calculations had implied that, even if it started at red heat, the Earth would have dropped to its present temperature in a few tens of millions of years. Armed with the knowledge of a new heat source, scientists realized that the Earth would be much older, and that its core was still sufficiently hot to be liquid.

By 1915, after having published a first article in 1912, Alfred Wegener was making serious arguments for the idea of continental drift in the first edition of *The Origin of Continents and Oceans*. In that book (re-issued in four successive editions up to the final one in 1936), he noted how the east coast of South America and the west coast of Africa looked as if they were once attached. Wegener was not the first to note this (Abraham Ortelius, Antonio Snider-Pellegrini, Eduard Suess, Roberto Mantovani and Frank Bursley Taylor preceded him just to mention a few), but he was the first to marshal significant fossil and paleo-topographical and climatological evidence to support this simple observation (and was supported in this by researchers such as Alex du Toit). Furthermore, when the rock strata of the margins of separate continents are very similar it suggests that these rocks were formed in the same way, implying that they were joined initially. For instance, parts of Scotland and Ireland contain rocks very similar to those found in Newfoundland and New Brunswick. Furthermore, the Caledonian Mountains of Europe and parts of the Appalachian Mountains of North America are very similar in structure and lithology.

However, his ideas were not taken seriously by many geologists, who pointed out that there was no apparent mechanism for continental drift. Specifically, they did not see how continental rock could plow through the much denser rock that makes up oceanic crust. Wegener could not explain the force that drove continental drift, and his vindication did not come until after his death in 1930.

Floating Continents, Paleomagnetism, and Seismicity Zones

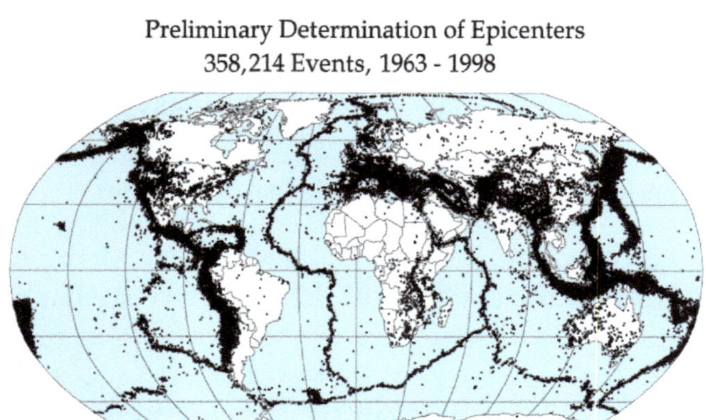

Global earthquake epicenters, 1963–1998

As it was observed early that although granite existed on continents, seafloor seemed to be composed of denser basalt, the prevailing concept during the first half of the twentieth century was that there were two types of crust, named "sial" (continental type crust) and "sima" (oceanic type crust). Furthermore, it was supposed that a static shell of strata was present under the continents. It therefore looked apparent that a layer of basalt (sial) underlies the continental rocks.

However, based on abnormalities in plumb line deflection by the Andes in Peru, Pierre Bouguer had deduced that less-dense mountains must have a downward projection into the denser layer underneath. The concept that mountains had "roots" was confirmed by George B. Airy a hundred years later, during study of Himalayan gravitation, and seismic studies detected corresponding density variations. Therefore, by the mid-1950s, the question remained unresolved as to whether mountain roots were clenched in surrounding basalt or were floating on it like an iceberg.

During the 20th century, improvements in and greater use of seismic instruments such as seismographs enabled scientists to learn that earthquakes tend to be concentrated in specific areas, most notably along the oceanic trenches and spreading ridges. By the late 1920s, seismologists were beginning to identify several prominent earthquake zones parallel to the trenches that typically were inclined 40–60° from the horizontal and extended several hundred kilometers into the Earth. These zones later became known as Wadati-Benioff zones, or simply Benioff zones, in honor of the seismologists who first recognized them, Kiyoo Wadati of Japan and Hugo Benioff of the United States. The study of global seismicity greatly advanced in the 1960s with the establishment of the Worldwide Standardized Seismograph Network (WWSSN) to monitor the compliance of the 1963 treaty banning above-ground testing of nuclear weapons. The much improved data from the WWSSN instruments allowed seismologists to map precisely the zones of earthquake concentration worldwide.

Meanwhile, debates developed around the phenomena of polar wander. Since the early debates of continental drift, scientists had discussed and used evidence that polar drift had occurred because continents seemed to have moved through different climatic zones during the past. Furthermore, paleomagnetic data had shown that the magnetic pole had also shifted during time. Reasoning in an opposite way, the continents might have shifted and rotated, while the pole remained relatively fixed. The first time the evidence of magnetic polar wander was used to support the movements of continents was in a paper by Keith Runcorn in 1956, and successive papers by him and his students Ted Irving (who was actually the first to be convinced of the fact that paleomagnetism supported continental drift) and Ken Creer.

This was immediately followed by a symposium in Tasmania in March 1956. In this symposium, the evidence was used in the theory of an expansion of the global crust. In this hypothesis the shifting of the continents can be simply explained by a large increase in size of the Earth since its formation. However, this was unsatisfactory because its supporters could offer no convincing mechanism to produce a significant expansion of the Earth. Certainly there is no evidence that the moon has expanded in the past 3 billion years; other work would soon show that the evidence was equally in support of continental drift on a globe with a stable radius.

During the thirties up to the late fifties, works by Vening-Meinesz, Holmes, Umbgrove, and numerous others outlined concepts that were close or nearly identical to modern plate tectonics theory. In particular, the English geologist Arthur Holmes proposed in 1920 that plate junctions might lie beneath the sea, and in 1928 that convection currents within the mantle might be the driving force. Often, these contributions are forgotten because:

- At the time, continental drift was not accepted.
- Some of these ideas were discussed in the context of abandoned fixistic ideas of a deforming globe without continental drift or an expanding Earth.
- They were published during an episode of extreme political and economic instability that hampered scientific communication.
- Many were published by European scientists and at first not mentioned or given little credit in the papers on sea floor spreading published by the American researchers in the 1960s.

Mid-oceanic Ridge Spreading and Convection

In 1947, a team of scientists led by Maurice Ewing utilizing the Woods Hole Oceanographic Institution's research vessel *Atlantis* and an array of instruments, confirmed the existence of a rise in the central Atlantic Ocean, and found that the floor of the seabed beneath the layer of sediments consisted of basalt, not the granite which is the main constituent of continents. They also found that the oceanic crust was much thinner than continental crust. All these new findings raised important and intriguing questions.

The new data that had been collected on the ocean basins also showed particular characteristics regarding the bathymetry. One of the major outcomes of these datasets was that all along the globe, a system of mid-oceanic ridges was detected. An important conclusion was that along this system, new ocean floor was being created, which led to the concept of the "Great Global Rift". This was described in the crucial paper of Bruce Heezen (1960), which would trigger a real revolution in thinking. A profound consequence of seafloor spreading is that new crust was, and still is, being continually created along the oceanic ridges. Therefore, Heezen advocated the so-called "expanding Earth" hypothesis of S. Warren Carey. So, still the question remained: how can new crust be continuously added along the oceanic ridges without increasing the size of the Earth? In reality, this question had been solved already by numerous scientists during the forties and the fifties, like Arthur Holmes, Vening-Meinesz, Coates and many others: The crust in excess disappeared along what were called the oceanic trenches, where so-called "subduction" occurred. Therefore, when various scientists during the early sixties started to reason on the data at their disposal regarding the ocean floor, the pieces of the theory quickly fell into place.

The question particularly intrigued Harry Hammond Hess, a Princeton University geologist and a Naval Reserve Rear Admiral, and Robert S. Dietz, a scientist with the U.S. Coast and Geodetic Survey who first coined the term *seafloor spreading*. Dietz and Hess (the former published the same idea one year earlier in *Nature*, but priority belongs to Hess who had already distributed an unpublished manuscript of his 1962 article by 1960) were among the small handful who really understood the broad implications of sea floor spreading and how it would eventually agree with the, at that time, unconventional and unaccepted ideas of continental drift and the elegant and mobilistic models proposed by previous workers like Holmes.

In the same year, Robert R. Coats of the U.S. Geological Survey described the main features of island arc subduction in the Aleutian Islands. His paper, though little noted (and even ridiculed) at the time, has since been called "seminal" and "prescient". In reality, it actually shows that the work by the European scientists on island arcs and mountain belts performed and published during the 1930s up until the 1950s was applied and appreciated also in the United States.

If the Earth's crust was expanding along the oceanic ridges, Hess and Dietz reasoned like Holmes and others before them, it must be shrinking elsewhere. Hess followed Heezen, suggesting that new oceanic crust continuously spreads away from the ridges in a conveyor belt–like motion. And, using the mobilistic concepts developed before, he correctly concluded that many millions of years later, the oceanic crust eventually descends along the continental margins where oceanic trenches – very deep, narrow canyons – are formed, e.g. along the rim of the Pacific Ocean basin. The important step Hess made was that convection currents would be the driving force in this process, arriving at the same conclusions as Holmes had decades before with the only difference that the thinning of the ocean crust was performed using Heezen's mechanism of spreading along the ridges. Hess therefore concluded that the Atlantic Ocean was expanding while the Pacific Ocean was shrinking. As old oceanic crust is "consumed" in the trenches (like Holmes and others, he thought this was done by thickening of the continental lithosphere, not, as now understood, by underthrusting at a larger scale of the oceanic crust itself into the mantle), new magma rises and erupts along the spreading ridges to form new crust. In effect, the ocean basins are perpetually being "recycled," with the creation of new crust and the destruction of old oceanic lithosphere occurring simultaneously. Thus, the new mobilistic concepts neatly explained why the Earth does not get bigger with sea floor spreading, why there is so little sediment accumulation on the ocean floor, and why oceanic rocks are much younger than continental rocks.

Magnetic Striping

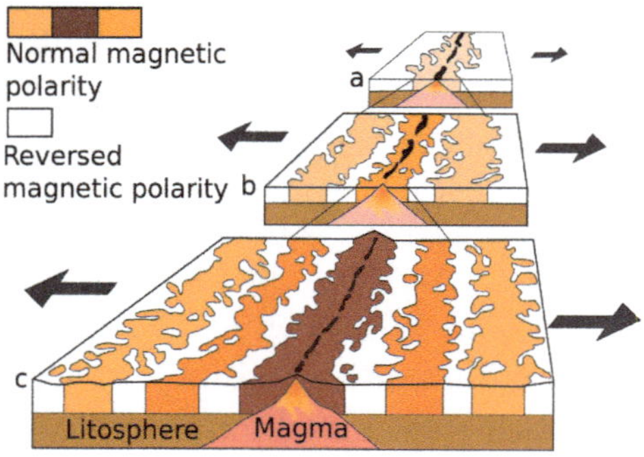

Seafloor magnetic striping.

A demonstration of magnetic striping. (The darker the color is, the closer it is to normal polarity)

Beginning in the 1950s, scientists like Victor Vacquier, using magnetic instruments (magnetometers) adapted from airborne devices developed during World War II to detect submarines, began recognizing odd magnetic variations across the ocean floor. This finding, though unexpected, was not entirely surprising because it was known that basalt—the iron-rich, volcanic rock making up the ocean floor—contains a strongly magnetic mineral (magnetite) and can locally distort compass readings. This distortion was recognized by Icelandic mariners as early as the late 18th century. More important, because the presence of magnetite gives the basalt measurable magnetic properties, these newly discovered magnetic variations provided another means to study the deep ocean floor. When newly formed rock cools, such magnetic materials recorded the Earth's magnetic field at the time.

As more and more of the seafloor was mapped during the 1950s, the magnetic variations turned out not to be random or isolated occurrences, but instead revealed recognizable patterns. When these magnetic patterns were mapped over a wide region, the ocean floor showed a zebra-like pattern: one stripe with normal polarity and the adjoining stripe with reversed polarity. The overall pattern, defined by these alternating bands of normally and reversely polarized rock, became known as magnetic striping, and was published by Ron G. Mason and co-workers in 1961, who did not find, though, an explanation for these data in terms of sea floor spreading, like Vine, Matthews and Morley a few years later.

The discovery of magnetic striping called for an explanation. In the early 1960s scientists such as Heezen, Hess and Dietz had begun to theorise that mid-ocean ridges mark structurally weak zones where the ocean floor was being ripped in two lengthwise along the ridge crest. New magma from deep within the Earth rises easily through these weak zones and eventually erupts along the crest of the ridges to create new oceanic crust. This process, at first denominated the "conveyer belt hypothesis" and later called seafloor spreading, operating over many millions of years continues to form new ocean floor all across the 50,000 km long system of mid-ocean ridges.

Only four years after the maps with the "zebra pattern" of magnetic stripes were published, the link between sea floor spreading and these patterns was correctly placed, independently by Lawrence Morley, and by Fred Vine and Drummond Matthews, in 1963, now called the Vine-Matthews-Morley hypothesis. This hypothesis linked these patterns to geomagnetic reversals and was supported by several lines of evidence:

1. the stripes are symmetrical around the crests of the mid-ocean ridges; at or near the crest of the ridge, the rocks are very young, and they become progressively older away from the ridge crest;
2. the youngest rocks at the ridge crest always have present-day (normal) polarity;
3. stripes of rock parallel to the ridge crest alternate in magnetic polarity (normal-reversed-normal, etc.), suggesting that they were formed during different epochs documenting the (already known from independent studies) normal and reversal episodes of the Earth's magnetic field.

By explaining both the zebra-like magnetic striping and the construction of the mid-ocean ridge system, the seafloor spreading hypothesis (SFS) quickly gained converts and represented another major advance in the development of the plate-tectonics theory. Furthermore, the oceanic crust now came to be appreciated as a natural "tape recording" of the history of the geomagnetic field reversals (GMFR) of the Earth's magnetic field. Today, extensive studies are dedicated to the calibration of the normal-reversal patterns in the oceanic crust on one hand and known timescales derived from the dating of basalt layers in sedimentary sequences (magnetostratigraphy) on the other, to arrive at estimates of past spreading rates and plate reconstructions.

Definition and Refining of the Theory

After all these considerations, Plate Tectonics (or, as it was initially called "New Global Tectonics") became quickly accepted in the scientific world, and numerous papers followed that defined the concepts:

- In 1965, Tuzo Wilson who had been a promotor of the sea floor spreading hypothesis and continental drift from the very beginning added the concept of transform faults to the model, completing the classes of fault types necessary to make the mobility of the plates on the globe work out.

- A symposium on continental drift was held at the Royal Society of London in 1965 which must be regarded as the official start of the acceptance of plate tectonics by the scientific community, and which abstracts are issued as Blacket, Bullard & Runcorn (1965). In this symposium, Edward Bullard and co-workers showed with a computer calculation how the continents along both sides of the Atlantic would best fit to close the ocean, which became known as the famous "Bullard's Fit".
- In 1966 Wilson published the paper that referred to previous plate tectonic reconstructions, introducing the concept of what is now known as the "Wilson Cycle".
- In 1967, at the American Geophysical Union's meeting, W. Jason Morgan proposed that the Earth's surface consists of 12 rigid plates that move relative to each other.
- Two months later, Xavier Le Pichon published a complete model based on 6 major plates with their relative motions, which marked the final acceptance by the scientific community of plate tectonics.
- In the same year, McKenzie and Parker independently presented a model similar to Morgan's using translations and rotations on a sphere to define the plate motions.

Implications for Biogeography

Continental drift theory helps biogeographers to explain the disjunct biogeographic distribution of present-day life found on different continents but having similar ancestors. In particular, it explains the Gondwanan distribution of ratites and the Antarctic flora.

Plate Reconstruction

Reconstruction is used to establish past (and future) plate configurations, helping determine the shape and make-up of ancient supercontinents and providing a basis for paleogeography.

Defining Plate Boundaries

Current plate boundaries are defined by their seismicity. Past plate boundaries within existing plates are identified from a variety of evidence, such as the presence of ophiolites that are indicative of vanished oceans.

Past Plate Motions

Tectonic motion first began around three billion years ago.

Various types of quantitative and semi-quantitative information are available to constrain past plate motions. The geometric fit between continents, such as between west Africa and South America is still an important part of plate reconstruction. Magnetic stripe patterns provide a reliable guide to relative plate motions going back into the Jurassic period. The tracks of hotspots give absolute reconstructions, but these are only available back to the Cretaceous. Older reconstructions rely mainly on paleomagnetic pole data, although these only constrain the latitude and rotation, but not the longitude. Combining poles of different ages in a particular plate to produce apparent polar wander paths provides a method for comparing the motions of different plates through time. Additional evidence comes from the distribution of certain sedimentary rock types, faunal provinces shown by particular fossil groups, and the position of orogenic belts.

Formation and Break-up of Continents

The movement of plates has caused the formation and break-up of continents over time, including occasional formation of a supercontinent that contains most or all of the continents. The supercontinent Columbia or Nuna formed during a period of 2,000 to 1,800 million years ago and broke up about 1,500 to 1,300 million years ago. The supercontinent Rodinia is thought to have formed about 1 billion years ago and to have embodied most or all of Earth's continents, and broken up into eight continents around 600 million years ago. The eight continents later re-assembled into another supercontinent called Pangaea; Pangaea broke up into Laurasia (which became North America and Eurasia) and Gondwana (which became the remaining continents).

The Himalayas, the world's tallest mountain range, are assumed to have been formed by the collision of two major plates. Before uplift, they were covered by the Tethys Ocean.

Current Plates

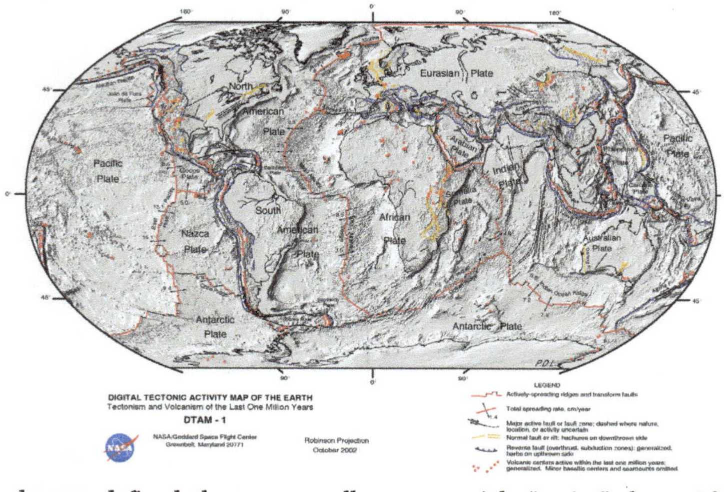

Depending on how they are defined, there are usually seven or eight "major" plates: African, Antarctic, Eurasian, North American, South American, Pacific, and Indo-Australian. The latter is sometimes subdivided into the Indian and Australian plates.

There are dozens of smaller plates, the seven largest of which are the Arabian, Caribbean, Juan de Fuca, Cocos, Nazca, Philippine Sea and Scotia.

The current motion of the tectonic plates is today determined by remote sensing satellite data sets, calibrated with ground station measurements.

Other Celestial Bodies (Planets, Moons)

The appearance of plate tectonics on terrestrial planets is related to planetary mass, with more massive planets than Earth expected to exhibit plate tectonics. Earth may be a borderline case, owing its tectonic activity to abundant water (silica and water form a deep eutectic.)

Venus

Venus shows no evidence of active plate tectonics. There is debatable evidence of active tectonics in the planet's distant past; however, events taking place since then (such as the plausible and generally accepted hypothesis that the Venusian lithosphere has thickened greatly over the course of several hundred million years) has made constraining the course of its geologic record difficult. However, the numerous well-preserved impact craters have been utilized as a dating method to approximately date the Venusian surface (since there are thus far no known samples of Venusian rock to be dated by more reliable methods). Dates derived are dominantly in the range 500 to 750 million years ago, although ages of up to 1,200 million years ago have been calculated. This research has led to the fairly well accepted hypothesis that Venus has undergone an essentially complete volcanic resurfacing at least once in its distant past, with the last event taking place approximately within the range of estimated surface ages. While the mechanism of such an impressive thermal event remains a debated issue in Venusian geosciences, some scientists are advocates of processes involving plate motion to some extent.

One explanation for Venus's lack of plate tectonics is that on Venus temperatures are too high for significant water to be present. The Earth's crust is soaked with water, and water plays an important role in the development of shear zones. Plate tectonics requires weak surfaces in the crust along which crustal slices can move, and it may well be that such weakening never took place on Venus because of the absence of water. However, some researchers remain convinced that plate tectonics is or was once active on this planet.

Mars

Mars is considerably smaller than Earth and Venus, and there is evidence for ice on its surface and in its crust.

In the 1990s, it was proposed that Martian Crustal Dichotomy was created by plate tectonic processes. Scientists today disagree, and think that it was created either by upwelling within the Martian mantle that thickened the crust of the Southern Highlands and formed Tharsis or by a giant impact that excavated the Northern Lowlands.

Valles Marineris may be a tectonic boundary.

Observations made of the magnetic field of Mars by the *Mars Global Surveyor* spacecraft in 1999 showed patterns of magnetic striping discovered on this planet. Some scientists interpreted these as requiring plate tectonic processes, such as seafloor spreading. However, their data fail a "magnetic reversal test", which is used to see if they were formed by flipping polarities of a global magnetic field.

Icy Satellites

Some of the satellites of Jupiter have features that may be related to plate-tectonic style deformation, although the materials and specific mechanisms may be different from plate-tectonic activity on Earth. On 8 September 2014, NASA reported finding evidence of plate tectonics on Europa, a satellite of Jupiter—the first sign of such geological activity on another world other than Earth.

Titan, the largest moon of Saturn, was reported to show tectonic activity in images taken by the *Huygens* probe, which landed on Titan on January 14, 2005.

Exoplanets

On Earth-sized planets, plate tectonics is more likely if there are oceans of water; however, in 2007, two independent teams of researchers came to opposing conclusions about the likelihood of plate tectonics on larger super-earths with one team saying that plate tectonics would be episodic or stagnant and the other team saying that plate tectonics is very likely on super-earths even if the planet is dry.

Transform Fault

A transform fault or transform boundary (also known as a conservative plate boundary, since these faults neither create nor destroy lithosphere), is a type of fault whose relative motion is predominantly horizontal, in either a sinistral (left lateral) or dextral (right lateral) direction. Furthermore, transform faults end abruptly and are connected on both ends to other faults, ridges, or subduction zones. While most transform faults are hidden in the deep oceans where they form a series of short zigzags accommodating seafloor spreading, the best-known (and most destructive) are those on land at the margins of tectonic plates. Transform faults are the only type of strike-slip fault that can be classified as a plate boundary.

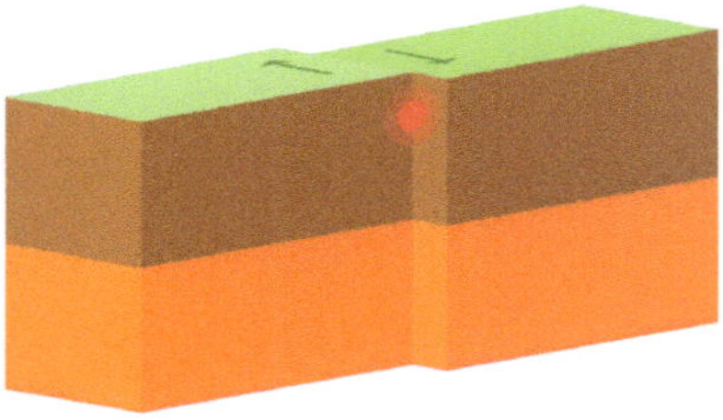

Diagram showing a transform fault with two plates moving in opposite directions

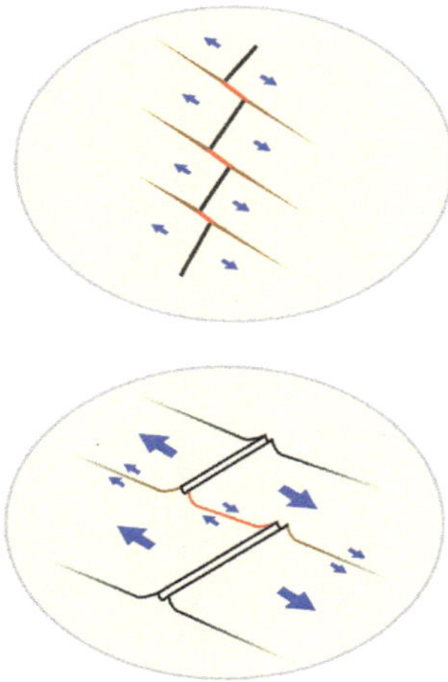

Transform fault (the red lines)

Background

John Tuzo Wilson recognized that the offsets of oceanic ridges by faults do not follow the classical pattern of an offset fence or geological marker in Reid's rebound theory of faulting, from which the sense of slip is derived. The new class of faults, called transform faults, produce slip in the opposite direction from what one would surmise from the standard interpretation of an offset geological feature. Slip along transform faults does not increase the distance between the ridges it separates; the distance remains constant in earthquakes because the ridges are spreading centers. This hypothesis was confirmed in a study of the fault plane solutions that showed the slip on transform faults points in the opposite direction than classical interpretation would suggest.

Difference between Transform and Transcurrent Faults

Transform faults are closely related to transcurrent faults, and are commonly confused. Both types of faults are strike-slip or side-to-side in movement, however transform faults end at the junction of another plate boundary or fault type, while transcurrent faults die out without a junction. In addition, transform faults have equal deformation across the entire fault line, while transcurrent faults have greater displacement in the middle of the fault zone and less on the margins. Finally, transform faults can form a tectonic plate boundary, while transcurrent faults cannot.

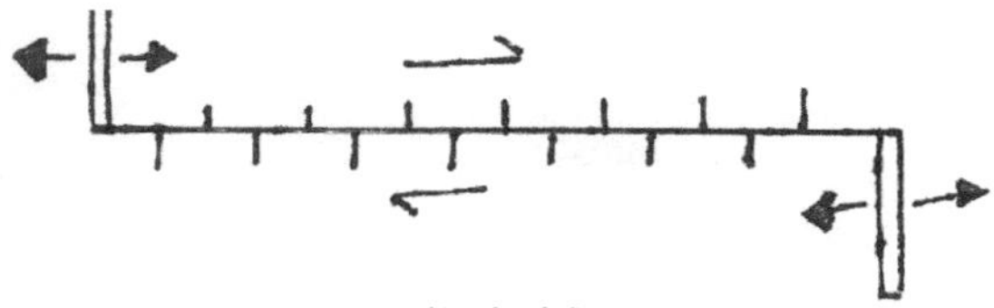

A transform fault
Linked on both ends by another tectonic boundary structure. As well as having equal deformation along the entire length of the fault line.

Transform fault

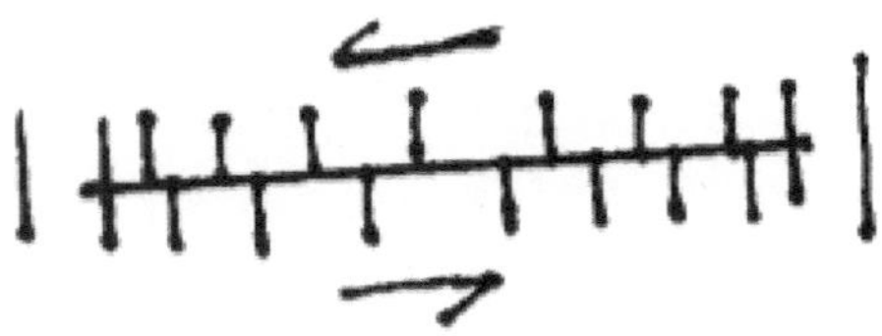

A transcurrent fault with unequal deformation.

Transcurrent fault

Mechanics

The effect of a fault is to relieve strain, which can be caused by compression, extension, or lateral stress in the rock layers at the surface or deep in the Earth's subsurface. Transform faults specifically relieve strain by transporting the strain between ridges or subduction zones. Transform faults also act as the plane of weakness allowing for the splitting in rift zones.

Examples

Transform faults are commonly found linking segments of mid-oceanic ridges or spreading centres. These mid-oceanic ridges are where new seafloor is constantly created through the upwelling of new basaltic magma. With new seafloor being pushed and pulled out, the older seafloor slowly slides away from the mid-oceanic ridges toward the continents. Although separated only by tens of kilometers, this separation between segments of the ridges causes portions of the seafloor to push past each other in opposing directions. This lateral movement of seafloors past each other is where transform faults are currently active.

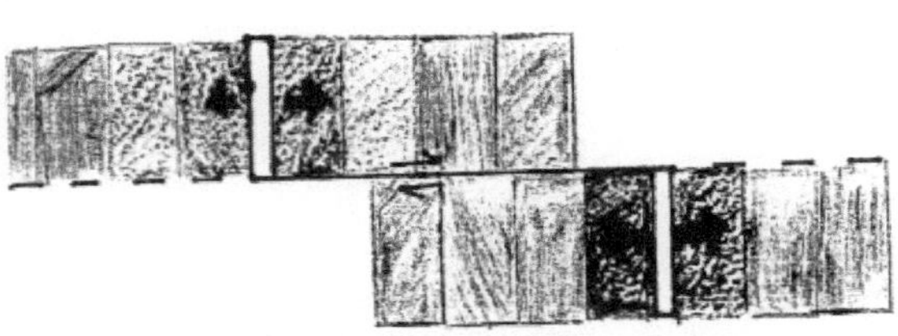

The blocks represent different magnetic strips in the sea floor moving in opposite directions to each other. The solid line represents the active transform fault and the dashed lines represent the fracture zones (inactive areas).

Spreading center and strips

Transform faults move differently than a strike-slip fault at the mid-oceanic ridge. Instead of the ridges moving away from each other, like other strike-slip faults, transform fault ridges will stay in the same fixed location, and the new ocean seafloor being created at the ridges is pushed away from the ridge. Evidence of this can be found in paleomagnetic striping on the seafloor.

A paper written by Gerya theorizes that the creation of the transform faults between the ridges of the mid-oceanic ridge is attributed to rotated and stretched sections of the mid-oceanic ridge. This occurs over a long period of time with the spreading center or ridge slowly deforming from a straight line to a curved line. Finally, fracturing along these planes forms transform faults. As this takes place, the fault changes from a normal fault with extensional stress to a strike slip fault with lateral stress. In the study done by Bonatti & Crane, peridotite and gabbro rocks were discovered in the edges of the transform ridges. These rocks are created deep inside the Earth's mantle and then rapidly exhumed to the surface. This evidence helps to prove that new seafloor is being created at the mid-oceanic ridges and further supports the theory of plate tectonics.

As previously stated, active transform faults are between two tectonic structures or faults. Fracture zones represent the previously active transform fault lines, which have since passed the active transform zone and are being pushed toward the continents. These elevated ridges on the ocean floor can be traced for hundreds of miles and in some cases even from one continent across an ocean to the other continent.

The most prominent examples of the mid-oceanic ridge transform zones are located in Atlantic Ocean between South America and Africa. Known as the St. Paul, Romanche, Chain, and Ascension fracture zones, these areas have with deep, easily identifiable transform faults and ridges. Other locations include: the East Pacific Ridge located in the South Eastern Pacific Ocean, which meets up with San Andreas Fault to the North.

The Southern Alps rise dramatically beside the Alpine Fault on New Zealand's West Coast. About 500 kilometres (300 mi) long; northwest at top.

Transform faults are not limited to oceanic crust and spreading centers; many transform faults are located on continental margins. The best example is the San Andreas Fault on the Pacific coast of the United States. The San Andreas Fault links the East Pacific Rise off of the West coast of Mexico (Gulf of California) to the Mendocino Triple Junction (Part of the Juan de Fuca plate) located off the coast of the North Western United States making it a ridge-to-transform style transform fault. The formation of the San Andreas Fault system occurred fairly recently during the Oligocene Period between 34 million and 24 million years ago. During this period, the Farallon plate, followed by the Pacific plate, collided into the North American plate. The collision led to the subduction of the Farallon plate underneath the North American plate. Once the spreading center separating the Pacific and Farallon plate was subducted underneath the North American plate, the San Andreas Continental Transform Fault system was created.

Other examples include:

- Middle East's Dead Sea Transform fault
- New Zealand's Alpine Fault
- Pakistan's Chaman Fault
- Turkey's North Anatolian Fault
- North America's Queen Charlotte Fault

Transform Fault Types

In his groundbreaking work on transform fault systems, Tuzo Wilson said that transform faults must be connected to other faults or tectonic plate boundaries on both ends; because of that requirement, transform faults can grow in length, keep a constant length, or decrease in length. These length changes are dependent on which type of faults or tectonic structures connect with the transform fault. With this in mind, Wilson described six types of transform faults:

Growing length faults: In situations where a transform fault links together a spreading center and the upper block of a subduction zone or when two upper blocks of subduction zones are linked the transform fault itself will grow in length.

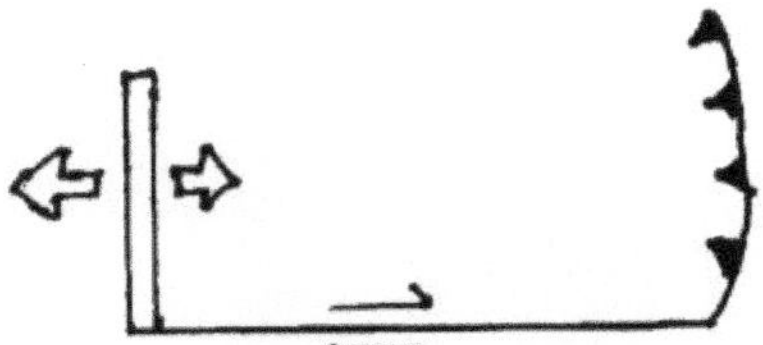

A spreading center (left) connected by a transform fault to the upper slab of a subduction zone

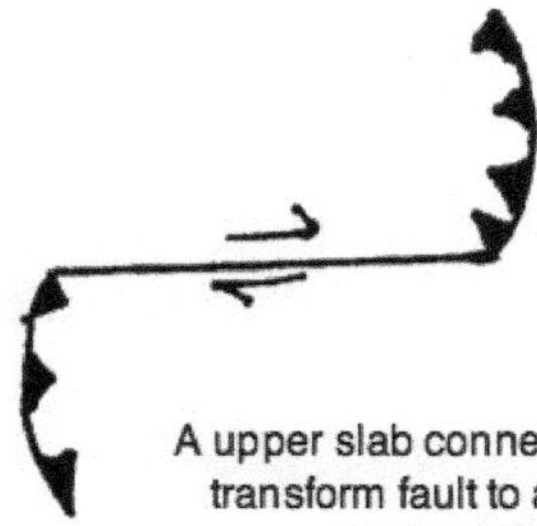

A upper slab connected by a transform fault to another upper slab.

Constant length faults: In other cases, transform faults will remain at a constant length. This consistency can be attributed to many different reasons. In the case of a ridge-to-ridge transforms, it is caused by the continuous growth by both ridges outward, canceling any change in length. The opposite occurs when a ridge linked to a subducting plate, where all the lithosphere (new sea floor) being created by the ridge is being subducted, or swallowed up, by the subduction zone. Finally, when two upper subduction plates are linked there is no change in length. This is due to the plates moving parallel with each other and no new lithosphere is being created to change that length.

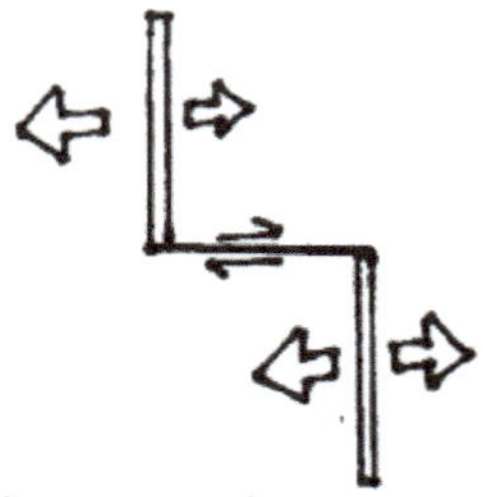

A transform fault connecting two mid-oceanic ridges.

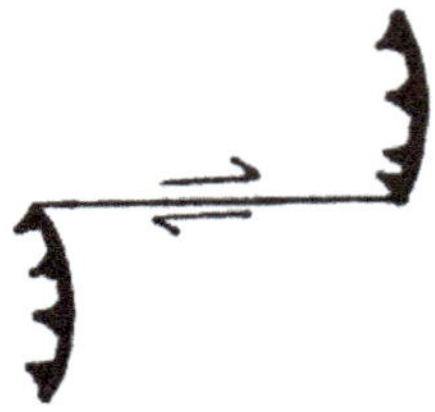

An upper slab of a subuction connected by a transform fault to descending slab of a subduction zone

A spreading center (left) connected by a transform fault to the descending slab of a subduction zone.

Decreasing length faults: In rare cases, transform faults can shrink in length. These occur when two descending subduction plates are linked by a transform fault. In time as the plates are subducted, the transform fault will decrease in length until the transform fault disappears completely, leaving only two subduction zones facing in opposite directions.

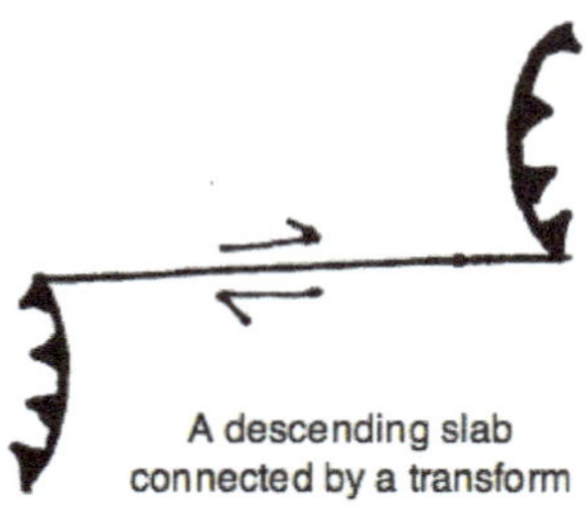

A descending slab connected by a transform fault to another descending slab.

Divergent Boundary

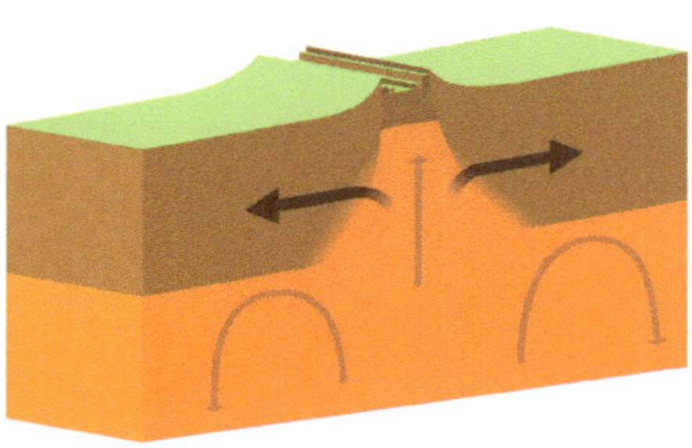

Continental-continental divergent/constructive boundary

In plate tectonics, a divergent boundary or divergent plate boundary (also known as a constructive boundary or an extensional boundary) is a linear feature that exists between two tectonic plates that are moving away from each other. Divergent boundaries within continents initially produce rifts which eventually become rift valleys. Most active divergent plate boundaries occur between oceanic plates and exist as mid-oceanic ridges. Divergent boundaries also form volcanic islands which occur when the plates move apart to produce gaps which molten lava rises to fill.

Current research indicates that complex convection within the Earth's mantle allows material to rise to the base of the lithosphere beneath each divergent plate boundary. This supplies the area with vast amounts of heat and a reduction in pressure that melts rock from the asthenosphere (or upper mantle) beneath the rift area forming large flood basalt or lava flows. Each eruption occurs in only a part of the plate boundary at any one time, but when it does occur, it fills in the opening gap as the two opposing plates move away from each other.

Over millions of years, tectonic plates may move many hundreds of kilometers away from both sides of a divergent plate boundary. Because of this, rocks closest to a boundary are younger than rocks further away on the same plate.

Description

Bridge across the Álfagjá rift valley in southwest Iceland, that is part of the boundary between the Eurasian and North American continental tectonic plates.

At divergent boundaries, two plates move apart from each other and the space that this creates is filled with new crustal material sourced from molten magma that forms below. The origin of new divergent boundaries at triple junctions is sometimes thought to be associated with the phenomenon known as hotspots. Here, exceedingly large convective cells bring very large quantities of hot asthenospheric material near the surface and the kinetic energy is thought to be sufficient to break apart the lithosphere. The hot spot which may have initiated the Mid-Atlantic Ridge system currently underlies Iceland which is widening at a rate of a few centimeters per year.

Divergent boundaries are typified in the oceanic lithosphere by the rifts of the oceanic ridge system, including the Mid-Atlantic Ridge and the East Pacific Rise, and in the continental lithosphere by rift valleys such as the famous East African Great Rift Valley. Divergent boundaries can create massive fault zones in the oceanic ridge system. Spreading is generally not uniform, so where spreading rates of adjacent ridge blocks are different, massive transform faults occur. These are the fracture zones, many bearing names, that are a major source of submarine earthquakes. A sea floor map will show a rather strange pattern of blocky structures that are separated by linear features perpendicular to the ridge axis. If one views the sea floor between the fracture zones as conveyor belts carrying the ridge on each side of the rift away from the spreading center the action becomes clear. Crest depths of the old ridges, parallel to the current spreading center, will be older and deeper... (from thermal contraction and subsidence).

It is at mid-ocean ridges that one of the key pieces of evidence forcing acceptance of the seafloor spreading hypothesis was found. Airborne geomagnetic surveys showed a strange pattern of symmetrical magnetic reversals on opposite sides of ridge centers. The pattern was far too regular to be coincidental as the widths of the opposing bands were too closely matched. Scientists had been studying polar reversals and the link was made by Lawrence W. Morley, Frederick John Vine and Drummond Hoyle Matthews in the Morley–Vine–Matthews hypothesis. The magnetic banding directly corresponds with the Earth's polar reversals. This was confirmed by measuring the ages of the rocks within each band. The banding furnishes a map in time and space of both spreading rate and polar reversals.

Various Divergent Boundary

Gakkel Ridge

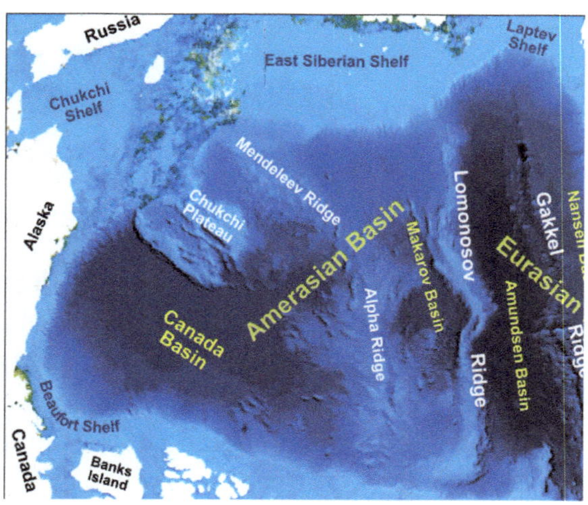

Main bathymetric/topographic features of the Arctic Ocean

The Gakkel Ridge (formerly known as the Nansen Cordillera and Arctic Mid-Ocean Ridge) is a mid-oceanic ridge, a divergent tectonic plate boundary between the North American Plate and the Eurasian Plate. It is located in the Arctic Ocean between Greenland and Siberia, and has a length of about 1,800 kilometers. Geologically, it connects the northern end of the Mid-Atlantic Ridge with the Laptev Sea Rift.

Pillow lava from the Gakkel Ridge

The existence and approximate location of the Gakkel Ridge were predicted by Soviet polar explorer Yakov Yakovlevich Gakkel, and confirmed on Soviet expeditions in the Arctic around 1950. The Ridge is named after him, and the name was recognized in April 1987 by SCUFN (under that body's old name, the Sub-Committee on Geographical Names and Nomenclature of Ocean Bottom Features).

The ridge is the slowest known spreading ridge on earth, with a rate of less than one centimeter per year. Until 1999, it was believed to be non-volcanic; that year, scientists operating from a nuclear submarine discovered active volcanos along it. In 2001 two research icebreakers, the German *Polarstern* and the American *Healy*, with several groups of scientists, cruised to the Gakkel Ridge to explore it and collect petrological samples. Among other discoveries, this expedition found evidence of hydrothermal vents. In 2007, Woods Hole Oceanographic Institution conducted the "Arctic Gakkel Vents Expedition" (AGAVE), which made some unanticipated discoveries, including the unconsolidated fragmented pyroclastic volcanic deposits that cover the axial valley of the ridge (whose area is greater than 10 km^2). These suggest volatile substances in concentrations ten times those in the magmas of normal mid-ocean ridges. Using "free-swimming" robotic submersibles on the Gakkel ridge, the AGAVE expedition also discovered what they called "bizarre 'mats' of microbial communities containing a half dozen or more new species".

East Pacific Rise

The East Pacific Rise is a mid-oceanic ridge, a divergent tectonic plate boundary located along the floor of the Pacific Ocean. It separates the Pacific Plate to the west from (north to south) the North American Plate, the Rivera Plate, the Cocos Plate, the Nazca Plate, and the Antarctic Plate. It runs from an undefined point near Antarctica in the south northward to its termination at the northern end of the Gulf of California in the Salton Sea basin in southern California.

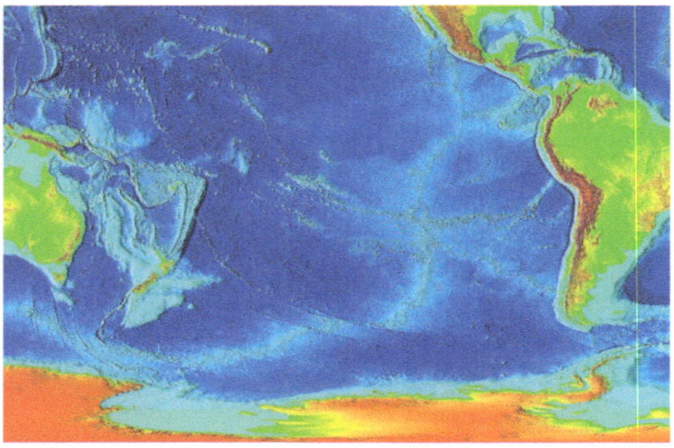

The East Pacific Rise, travelling South from the Gulf of California

Overview

East Pacific Rise, 21 degrees north. Base of "black smoker" chimney

The oceanic crust is moving away from the East Pacific Rise to either side at a rate of 7.5 cm/year. On the eastern side of the rise the eastward moving Cocos and Nazca plates meet the westward moving South American Plate and the North American Plate and are being subducted under them. The belt of volcanoes along the Andes and the arc of volcanoes through Central America and Mexico are the direct results of this collision. Due east of the Baja California Peninsula, the Rise is sometimes referred to as the Gulf of California Rift Zone. In this area, newly formed oceanic crust is intermingled with rifted continental crust originating from the North American Plate.

Near Easter Island, the East Pacific Rise meets the Chile Rise at the Easter Island and Juan Fernandez microplates, trending off to the east where it subducts under the South American Plate at the Peru–Chile Trench along the coast of southern Chile. The southern extension of the East Pacific Rise (called the Pacific-Antarctic Ridge) merges with the Southeast Indian Ridge at the Macquarie Triple Junction south of New Zealand.

Along the East Pacific Rise the hydrothermal vents called black smokers were first discovered and have been extensively studied. These vents are forming volcanogenic massive sulfide ore deposits on the ocean floor. Many strange deep-water creatures have been found here. The southern stretch of the East Pacific Rise is one of the fastest-spreading sections of the Earth's mid-ocean ridge system.

Explorer Ridge

The Explorer Ridge is a mid-ocean ridge, a divergent tectonic plate boundary located about 241 km (150 mi) west of Vancouver Island, British Columbia, Canada. It lies at the northern extremity of the Pacific spreading axis. To its east is the Explorer Plate, which together with the Juan de Fuca Plate and the Gorda Plate to its south, is what remains of the once-vast Farallon Plate which has been largely subducted under the North American Plate. The Explorer Ridge consists of one major segment, the Southern Explorer Ridge, and several smaller segments. It runs northward from the Queen Charlotte Triple Junction, a point where the Nootka Fault, the Sovanco Fracture Zone and the Juan de Fuca Ridge meet.

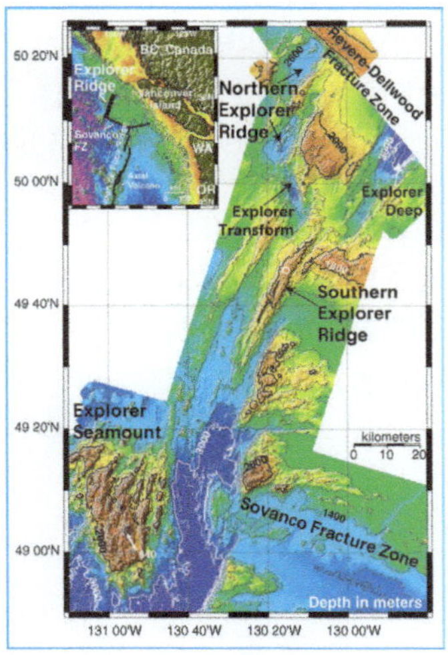

Bathymetry of Explorer Ridge area, collected between 1980 and 1984. Bathymetry is contoured at a 200-meter interval. The inset shows the ridge location. The open white circle represents the approximately location of the Magic Mountain vent site.

Pillow lavas and breccia overlain with slabby pieces of sulfide formed from hydrothermal venting on the east side of the Southern Explorer Ridge.

Bathymetry image showing the crest of Southern Explorer Ridge. Purple and dark blue colors indicate deepest depths.

Geology

This divergent boundary first formed about 5-7 million years ago when the northern end of the Juan de Fuca Plate broke off along the Nootka Fault to form the Explorer Plate. This apparently had some important ramifications for regional geologic evolution. When this change was completed, Cascade Arc volcanism from Northern California to southwestern British Columbia returned and the present-day Cascade Range and Olympic Mountains started to form.

The oceanic crust is moving away from the Explorer Ridge to either side. On the eastern side the eastward moving Explorer Plate is being subducted under the North American Plate. The belt of volcanoes along the Pacific Ranges are the direct results of this collision. The western side of the Explorer Ridge is associated with the northwest trending Pacific Plate which has formed the Queen Charlotte Fault, an active transform fault along the coast of British Columbia and southeast Alaska.

The Explorer Ridge is also seismically active. Most seismicity recorded in this area occurred on or near the Explorer Transform Fault Zone. The relatively shallow depth of the Southern Explorer Ridge (up to 1800 m) in comparison with most other segments of the northeast Pacific spreading centers suggests that there has been considerable volcanic activity along this segment in the past 100,000 years.

Notable Features along the Ridge

The Explorer Ridge includes a deep rift valley which runs along the axis of the ridge along nearly its entire length. This rift marks the actual boundary between adjacent tectonic plates, where magma from the mantle reaches the seafloor, erupting as lava and producing new crustal material for the plates.

Before 2002 Explorer Ridge was the least explored of the northeast Pacific spreading centers, even though it was known to have robust hydrothermal activity and is seismically active. Along the Southern Explorer Ridge lies a large hydrothermal vent area called Magic Mountain. It is an unusual hydrothermal site, with its off-axis location and relatively long-lived activity. The source of the hydrothermal fluid that fuels Magic Mountain probably rises along fault systems associated with a recent episode of rifting that, in turn, followed a massive outpouring of lava. These vents are forming seafloor massive sulfide deposits on the ocean floor. Many strange deep-water creatures have been found here.

Convergent Boundary

In plate tectonics, a convergent boundary, also known as a destructive plate boundary (because of subduction), is an actively deforming region where two (or more) tectonic plates or fragments of the lithosphere move toward one another and collide. As a result of pressure, friction, and plate material melting in the mantle, earthquakes and volcanoes are common near convergent boundaries. When two plates move towards one another, they form either a subduction zone or a continental collision. This depends on the nature of the plates involved. In a subduction zone, the subducting plate, which is normally a plate with oceanic crust, moves beneath the other plate, which can be made of either oceanic or continental crust. During collisions between two continental plates, large mountain ranges, such as the Himalayas are formed.

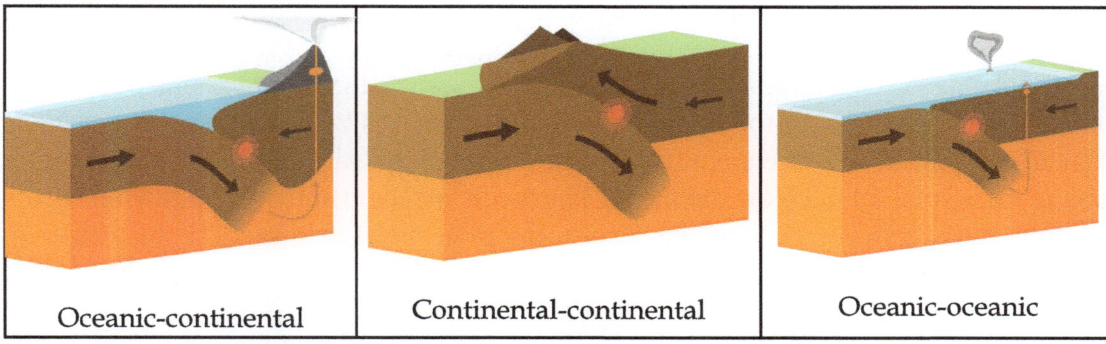

Descriptions

The nature of a convergent boundary depends on the type of plates that are colliding. Where a dense oceanic plate collides with a less-dense continental plate, the oceanic plate is typically thrust underneath because of the greater buoyancy of the continental lithosphere, forming a subduction zone. At the surface, the topographic expression is commonly an oceanic trench on the ocean side and a mountain range on the continental side. An example of a continental-oceanic subduction zone is the area along the western coast of South America where the oceanic Nazca Plate is being subducted beneath the continental South American Plate.

Surface volcanism (volcanoes at the ocean floor or the Earth's surface) typically appears above the melts which form directly above downgoing plates. There is still debate in the geologic community as to why this is, however, the general consensus from ongoing research suggests that the release of volatiles is the primary contributor. As the subducting plate descends, its temperature rises driving off volatiles (most importantly water) encased in the porous oceanic crust. As this water rises into the mantle of the overriding plate, it lowers the melting temperature of surrounding mantle, producing melts (magma) with large amounts of dissolved gases. These melts rise to the surface and are the source of some of the most explosive volcanism on Earth because of their high volumes of extremely pressurized gases (consider Mount St. Helens). The melts rise to the surface and cool, forming long chains of volcanoes inland from the continental shelf and parallel to it. The continental spine of western South America is dense with this type of volcanic mountain building from the subduction of the Nazca Plate. In North America the Cascade mountain range, extending north from California's Sierra

Nevada, is also of this tectonic type. Such volcanoes are characterized by alternating periods of quiet and episodic eruptions that start with explosive gas expulsion with fine particles of glassy volcanic ash and spongy cinders, followed by a rebuilding phase with hot magma. The entire Pacific Ocean boundary is surrounded by long stretches of volcanoes and is known collectively as the Pacific Ring of Fire.

Where two continental plates collide the plates either buckle and compress or (in some cases) one plate delves called subduction, under the other. Either action will create extensive mountain ranges. The most dramatic effect seen is where the northern margin of the Indian Plate is being thrust under a portion of the Eurasian Plate, lifting it and creating the Himalayas and the Tibetan Plateau beyond. It may have also pushed nearby parts of the Asian continent aside to the east.

When two plates with oceanic crust converge, they typically create an island arc as one plate is subducted below the other. The arc is formed from volcanoes which erupt through the overriding plate as the descending plate melts below it. The arc shape occurs because of the spherical surface of the earth (nick the peel of an orange with a knife and note the arc formed by the straight-edge of the knife). A deep oceanic trench is located in front of such arcs where the descending slab dips downward, such as the Mariana trench near the Mariana Islands.

Plates may collide at an oblique angle rather than head-on to each other (e.g. one plate moving north, the other moving south-east), and this may cause strike-slip faulting along the collision zone, in addition to subduction or compression.

Not all plate boundaries are easily defined. Some are broad belts whose movements are unclear to scientists. One example would be the Mediterranean-Alpine boundary, which involves two major plates (African and Eurasian) and several micro plates. The boundaries of the plates do not necessarily coincide with those of the continents. For instance, the North American Plate covers not only North America, but also far northeastern Siberia, plus a substantial portion of the Atlantic Ocean.

Convergent Margins

A subduction zone is formed at a convergent plate boundary when one or both of the tectonic plates is composed of oceanic crust. The denser plate, made of oceanic crust, is subducted underneath the less dense plate, which can be either continental or oceanic crust. When both of the plates are made of oceanic crust, convergence is associated with island arcs such as the Solomon Islands.

An oceanic trench is found where the denser plate is subducted underneath the other plate. There is water in the rocks of the oceanic plate (because they are underwater), and as this plate moves further down into the subduction zone, much of the water contained in the plate is squeezed out when the plate begins to subduct. However, the recrystallization of ocean floor rocks, such as Serpentine, which are unstable in the upper mantle, recrystallize into Olivine, causing dehydration through loss of hydroxyl groups. This addition of water to the mantle causes partial melting of the mantle, generating magma, which then rises, and which normally results in volcanoes. This normally happens at a certain depth, about 70 to 80 miles below the Earth's surface, and so volcanoes are formed fairly close to, but not right next to, the trench.

Some convergent margins have zones of active seafloor spreading behind the island arc, known as back-arc basins. When one plate is composed of oceanic lithosphere and the other is composed of continental lithosphere, the denser oceanic plate is subducted, often forming an orogenic belt and associated mountain range. This type of convergent boundary is similar to the Andes or the Cascade Range in North America.

When two plates containing continental crust collide, both are too light to subduct. In this case, a continent-continent collision occurs, creating especially large mountain ranges. The most spectacular example of this is the Himalayas.

When the subducting plate approaches the trench obliquely, the convergent plate boundary includes a major component of strike-slip faulting within the over-riding plate. The best example of this is the Sumatra convergent margin, where orthogonal convergence on the Sunda megathrust is occurring intermixed with movement on the Great Sumatran fault.

Plate Reconstruction

Plate reconstruction is the process of reconstructing the positions of tectonic plates relative to each other (relative motion) or to other reference frames, such as the earth's magnetic field or groups of hotspots, in the geological past. This helps determine the shape and make-up of ancient supercontinents and provides a basis for paleogeographic reconstructions.

Defining Plate Boundaries

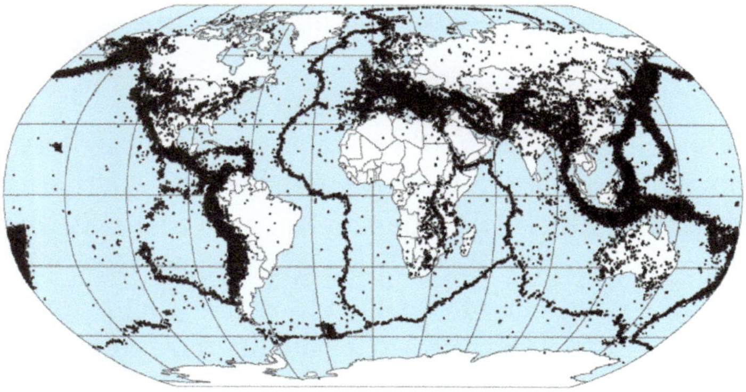

Earthquake epicenters 1963–98

An important part of reconstructing past plate configurations is to define the edges of areas of the lithosphere that have acted independently at some time in the past.

Present Plate Boundaries

Most present plate boundaries are easily identifiable from the pattern of recent seismicity. This is now backed up by the use of GPS data, to confirm the presence of significant relative movement between plates.

Past Plate Boundaries

Identifying past (but now inactive) plate boundaries within current plates is generally based on evidence for an ocean that has now closed up. The line where the ocean used to be is normally marked by pieces of the crust from that ocean, included in the collision zone, known as ophiolites. The line across which two plates became joined to form a single larger plate, is known as a suture.

In many orogenic belts, the collision is not just between two plates, but involves the sequential accretion of smaller terranes. Terranes are smaller pieces of continental crust that have been caught up in an orogeny, such as continental fragments or island arcs.

Reference Frames

Plate motions, both those observable now and in the past, are referred ideally to a reference frame that allows other plate motions to be calculated. For example, a central plate, such as the African plate, may have the motions of adjacent plates referred to it. By composition of reconstructions, additional plates can be reconstructed to the central plate. In turn, the reference plate may be reconstructed, together with the other plates, to another reference frame, such as the earth's magnetic field, as determined from paleomagnetic measurements of rocks of known age. A global hotspot reference frame has been postulated but there is now evidence that not all hotspots are necessarily fixed in their locations relative to one another or the earth's spin axis. However, there are groups of such hotspots that appear to be fixed within the constraints of available data, within particular mesoplates.

Euler Poles

The movement of a rigid body, such as a plate, on the surface of a sphere can be described as rotation about a fixed axis (relative to the chosen reference frame). This pole of rotation is known as an Euler pole. The movement of a plate is completely specified in terms of its Euler pole and the angular rate of rotation about the pole. Euler poles defined for current plate motions can be used to reconstruct plates in the recent past (few million years). At earlier stages of earth's history, new Euler poles need to be defined. It was suggested recently that paleomagnetic Euler poles derived from apparent polar wander path could potentially extend the upper age limit of marine geophysical observations and hot spot traces.

Estimating Past Plate Motions

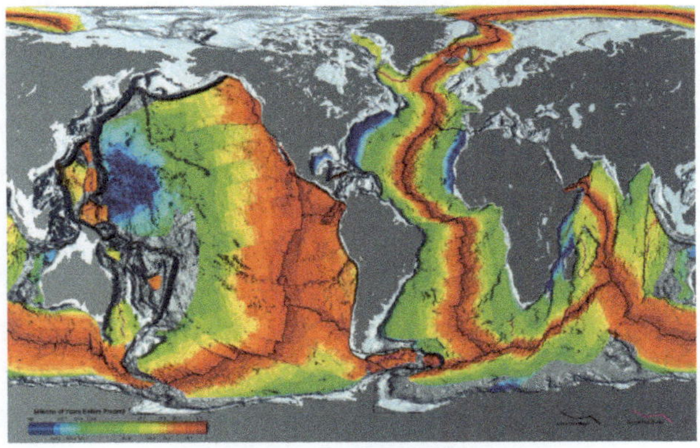

Ages of oceanic lithosphere

In order to move plates backward in time it is necessary to provide information on either relative or absolute positions of the plates being reconstructed such that an Euler pole can be calculated. These are quantitative methods of reconstruction.

Geometric Matching of Continental Borders

Certain fits between continents, particularly that between South America and Africa, were known long before the development of a theory that could adequately explain them. The reconstruction before Atlantic rifting by Bullard based on a least-squares fitting at the 500 fathom contour still provides the best match to paleomagnetic pole data for the two sides from the middle of Paleozoic to Late Triassic.

Plate Motion from Magnetic Stripes

Plate reconstructions in the recent geological past mainly use the pattern of magnetic stripes in oceanic crust to remove the effects of seafloor spreading. The individual stripes are dated from magnetostratigraphy so that their time of formation is known. Each stripe (and its mirror image) represents a plate boundary at a particular time in the past, allowing the two plates to be repositioned relative to one another. The oldest oceanic crust is of Jurassic age, providing a lower age limit of about 175 Ma for the use of such data. Reconstructions derived in this way are only relative.

Paleomagnetic Pole Data

Sampling

Paleomagnetic pole data is obtained by taking oriented samples of rock. Good quality poles have been recovered from many different rock types. In igneous rocks, the magnetic minerals have cooled through the curie point, typically providing good quality poles. Metamorphic rocks may contain new magnetic minerals but are not normally used to obtain poles due to their typically high magnetic anisotropy. Sedimentary rocks generally have a complex set of magnetic components such as derived clasts and mineral grains with their own inherited magnetization, primary magnetization acquired at the time of sedimentation and secondary magnetization related to subsequent diagenesis and authigenic mineral growth. Progressive demagnetization is used to discriminate between the components to identify the primary magnetization. The poles shown by the primary magnetization represent a record of their position relative to the magnetic pole at the time when the rock was formed. Suitable rocks of the right age are not present everywhere, so some pieces of the puzzle remain very unconstrained.

Good quality magnetic poles are added to the *Global Paleomagnetic Database*, which is accessible from the World Data Center A in the USA at Boulder, Colorado; and from the Norwegian Geological Survey in Trondheim.

Apparent Polar Wander Paths

Poles from different ages in a single area can be used to construct an apparent polar wander (APW) path. If paths from adjacent crustal fragments are identical, this is taken to indicate that there has been no relative movement between them during the period covered by the path. Divergence of APW paths indicates that the areas in question have acted independently in the past with the point of divergence marking the time at which they became joined.

Observed paleomagnetic poles show only the approximate latitude at the time but provide no constraint on the longitude. For this reason older plate reconstructions have much larger uncertainties associated with them. Paleomagnetic Euler rotations (rotation poles and angles) and the mantle structures, however, could provide clues to this longstanding challenge.

Hotspot Tracks

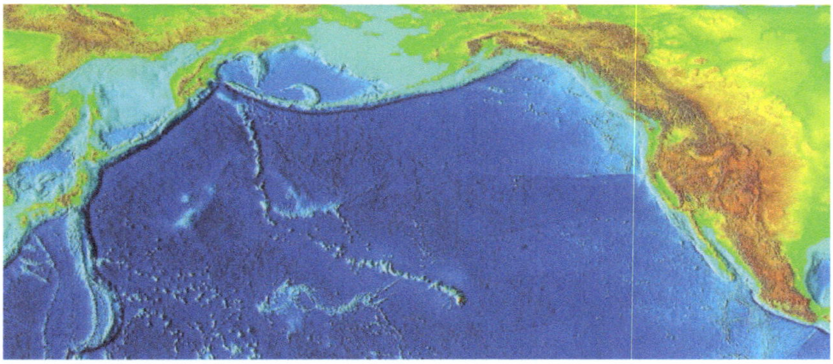

The Hawaiian-Emperor seamount chain

The presence of chains of volcanic islands and seamounts interpreted to have formed from fixed hotspots allows the plate on which they sit to be progressively restored so that a seamount is moved back over the hotspot at its time of formation. This method can be used back to the Early Cretaceous, the age of the oldest evidence for hotspot activity. This method gives an absolute reconstruction of both latitude and longitude, although before about 90 Ma there is evidence of relative motion between hotspot groups.

Slab Constraints

Once oceanic plates subduct in the lower mantle (slabs), they are assumed to sink in a near-vertical manner. With the help of seismic wave tomography, this can be used to constrain plate reconstructions at first order back to the Permian.

Apparent Polar Wander Paths Geometric Parameterizations

Paleomagnetic Euler poles derived by geometrizing apparent polar wander paths (APWPs) potentially allows constraining paleolongitudes from paleomagnetic data. This method could extend absolute plate motion reconstructions deeply into the geologic history as long as there are reliable APWPs.

Other Evidence for Past Plate Configurations

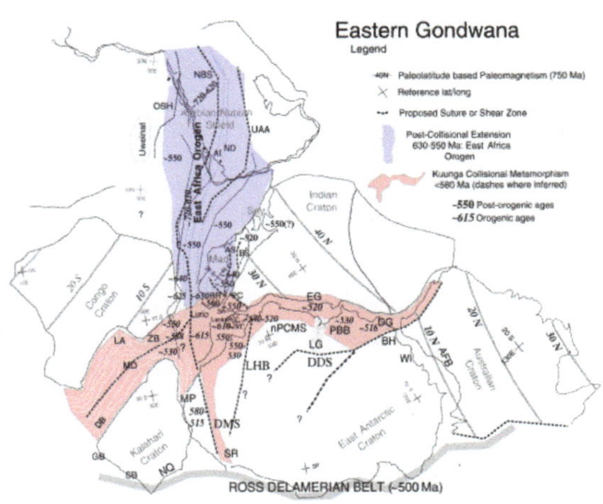

Reconstruction of eastern Gondwana showing position of orogenic belts

Some plate reconstructions are supported by other geological evidence, such as the distribution of sedimentary rock types, the position of orogenic belts and faunal provinces shown by particular fossils. These are semi-quantitative methods of reconstruction.

Sedimentary Rock Types

Some types of sedimentary rock are restricted to certain latitudinal belts. Glacial deposits for instance are generally confined to high latitudes, whereas evaporites are generally formed in the tropics.

Faunal Provinces

Oceans between continents provide barriers to plant and animal migration. Areas that have become separated tend to develop their own fauna and flora. This is particularly the case for plants and land animals but is also true for shallow water marine species, such as trilobites and brachiopods, although their planktonic larvae mean that they were able to migrate over smaller deep water areas. As oceans narrow before a collision occurs, the faunas start to become mixed again, providing supporting evidence for the closure and its timing.

Orogenic Belts

When supercontinents break up, older linear geological structures such as orogenic belts may be split between the resulting fragments. When a reconstruction effectively joins up orogenic belts of the same age of formation, this provides further support for the reconstruction's validity.

References

- Condie, K.C. (1997). Plate tectonics and crustal evolution (4th ed.). Butterworth-Heinemann. p. 282. ISBN 978-0-7506-3386-4. Retrieved 2010-02-21.
- Kearey, P.; Klepeis K.A. & Vine F.J. (2009). Global tectonics (3rd ed.). Wiley-Blackwell. p. 482. ISBN 978-1-4051-0777-8.
- Dyches, Preston; Brown, Dwayne; Buckley, Michael (8 September 2014). "Scientists Find Evidence of 'Diving' Tectonic Plates on Europa". NASA. Retrieved 8 September 2014.
- Wolpert, Stuart (August 9, 2012). "UCLA scientist discovers plate tectonics on Mars". Yin, An. UCLA. Retrieved August 13, 2012.
- Neith, Katie (April 15, 2011). "Caltech Researchers Use GPS Data to Model Effects of Tidal Loads on Earth's Surface". Caltech. Retrieved August 15, 2012.
- Butler, R.F. (1992). "Chapter 10 Applications to paleogeography" (PDF). Paleomagnetism:Magnetic domains to geologic terranes. Blackwell. Retrieved 22 February 2010.

Permissions

All chapters in this book are published with permission under the Creative Commons Attribution Share Alike License or equivalent. Every chapter published in this book has been scrutinized by our experts. Their significance has been extensively debated. The topics covered herein carry significant information for a comprehensive understanding. They may even be implemented as practical applications or may be referred to as a beginning point for further studies.

We would like to thank the editorial team for lending their expertise to make the book truly unique. They have played a crucial role in the development of this book. Without their invaluable contributions this book wouldn't have been possible. They have made vital efforts to compile up to date information on the varied aspects of this subject to make this book a valuable addition to the collection of many professionals and students.

This book was conceptualized with the vision of imparting up-to-date and integrated information in this field. To ensure the same, a matchless editorial board was set up. Every individual on the board went through rigorous rounds of assessment to prove their worth. After which they invested a large part of their time researching and compiling the most relevant data for our readers.

The editorial board has been involved in producing this book since its inception. They have spent rigorous hours researching and exploring the diverse topics which have resulted in the successful publishing of this book. They have passed on their knowledge of decades through this book. To expedite this challenging task, the publisher supported the team at every step. A small team of assistant editors was also appointed to further simplify the editing procedure and attain best results for the readers.

Apart from the editorial board, the designing team has also invested a significant amount of their time in understanding the subject and creating the most relevant covers. They scrutinized every image to scout for the most suitable representation of the subject and create an appropriate cover for the book.

The publishing team has been an ardent support to the editorial, designing and production team. Their endless efforts to recruit the best for this project, has resulted in the accomplishment of this book. They are a veteran in the field of academics and their pool of knowledge is as vast as their experience in printing. Their expertise and guidance has proved useful at every step. Their uncompromising quality standards have made this book an exceptional effort. Their encouragement from time to time has been an inspiration for everyone.

The publisher and the editorial board hope that this book will prove to be a valuable piece of knowledge for students, practitioners and scholars across the globe.

Index

U

V

W